The New Art of Old Public Sci Communication

I0037865

This book investigates the phenomenon of science communication events, as spectacles for legitimising and communicating science to the public. With attention to events such as 'Science Slam', where scientists are asked to present their knowledge in new ways and speak to an audience of laymen, the author examines the participants' use of stylistic devices borrowed from other events in order to address a diverse audience in a competitive environment. With attention to the performative appearance of scientists on stage and the manner in which contemporary public performing scientists present, problematise, and communicate knowledge, the author considers the justifications offered by participants in terms of legitimacy and expectations. Illustrating the crucial role of bodies, techniques, visuals, and objects in the communicative construction of (scientific) reality, *The New Art of Old Public Science Communication: The Science Slam* sheds new light on the construction of improved science communication. As such, it will appeal to social scientists with interests in science communication, the sociology of science and technology, and the sociology of knowledge.

Miira B. Hill is a postdoctoral researcher for Media and Communications at the ZeMKI (Center for Media, Communication and Information Research), University of Bremen, Germany.

Knowledge, Communication and Society

Series Editors

Bernt Schnettler
Universität Bayreuth, Germany
Hubert Knoblauch
Technische Universität Berlin, Germany
Michaela Pfadenhauer
University of Vienna, Austria
Alejandro Baer
University of Minnesota, USA

Knowledge, Communication, and Society: Contributions to the New Sociology of Knowledge seeks to revive the academic collaboration and debates between European and Anglo-Saxon currents of thought in the social sciences that characterised the middle of the last century, and provide a forum for the development of a new sociology of knowledge. A space for transatlantic discussion, it includes original works and translations of central works by contemporary European social scientists and is committed to an empirically grounded programme of developing social theory.

Titles in this series

Social Constructivism as Paradigm?
The Legacy of The Social Construction of Reality
Edited by Michaela Pfadenhauer and Hubert Knoblauch

Erving Goffman
From the Perspective of the New Sociology of Knowledge
Jürgen Raab

The Communicative Construction of Reality
Hubert Knoblauch

The New Art of Old Public Science Communication
The Science Slam
Miira B. Hill

For more information about this series, please visit: www.routledge.com/sociology/series/KCS

The New Art of Old Public Science Communication

The Science Slam

Miira B. Hill

Routledge
Taylor & Francis Group

LONDON AND NEW YORK

First published 2022
by Routledge
2 Park Square, Milton Park, Abingdon, Oxon OX14 4RN

and by Routledge
605 Third Avenue, New York, NY 10158

Routledge is an imprint of the Taylor & Francis Group, an informa business

British Library Cataloguing-in-Publication Data
A catalogue record for this book is available from the British Library

Library of Congress Cataloging-in-Publication Data
A catalog record for this book has been requested

ISBN: 978-1-032-00079-4 (hbk)
ISBN: 978-1-032-00080-0 (pbk)
ISBN: 978-1-003-17263-5 (ebk)

DOI: 10.4324/9781003172635

Typeset in Times New Roman
by Apex CoVantage, LLC

To Lilia, Leon and Benjamin

Contents

Illustrations

Figures

Images

Tables

Acknowledgements

This book was possible because the DFG supported me in the process of carrying out this research project as part of the graduate school, 'Innovation Society Today'. It is thanks to the painstaking care of Prof. Dr Hubert Knoblauch, and Prof. Dr Martin Reinhart that I could complete this work.

This book would not have been possible without the generous help from those in the Science Slam community, who have welcomed me to their gatherings, and even allowed me backstage at the events. My greatest thanks go to the founder of the Science Slam, Alexander Deppert, who invited me to a meeting of Science Slam organisers. I cannot express my gratitude enough for all the members of the Science Slam community who contributed to the project—there are too many to count, but I want to express my special acknowledgement to each and every one of them.

My debt to my colleagues in sociology, and science and technology studies is no less great. The critical, but productive, atmosphere at the Institute for Sociology at the Technical University Berlin was of great help during my research. The entire institute was extraordinarily cooperative. I thank Lilli Braunisch, Nina Amelung, Henning Mohr, Alexander Wentland, Gesche Joost, Bernt Schnettler, Michael Hutter, Lucy Suchman, Karin Knorr-Cetina, Emily York, Daniel Tödt, Victor Kempf, and Simone Jung: all scholars who contributed ideas and suggestions in response to papers, and in conversation. Thanks to Hannah Kropla, Inês Wilke, Jan Hussels, and Kamil Bembnista for assisting me in the last couple of years. I am also thankful to Yelena Gluzman and Charlotte Seddon, who revised the text. This list must be inconsiderably incomplete, and I want to apologise to, and at the same time thank, all those I omitted.

Also, thanks to Rosemarie Sanyang-Hill, Hubert Hill, and Fatou Sanyang for supporting me through seemingly endless periods of writing and rewriting, during which I proved to be an obsessed scientist! Lilia, Leon and Benjamin Hill are my greatest pride. Thanks for letting me be absent so often to do my research. I am thankful for the admirable teachers I had in school, including Klaus Peter Ifland. I want to thank Robert Hill for resettling our family in Germany from the United States. I am also grateful to Carla Felgentreff, Lisa Seibert, and Barbara Hache who offered helpful personal support while I was finishing the dissertation.

1   The Difficulty of
   Communicating Science
   to the Public

The communication of knowledge has an important role in contemporary, knowledge-based societies. The dominance of knowledge in various industries has resulted in a strong societal shift towards, and interest in, scientific knowledge. In these societies, where knowledge plays an important role, the pressure on science[1] to produce knowledge with non-scientific and economic relevance has grown. This development has several consequences. One such consequence is the need to communicate different types of knowledge to different areas of society. This leads naturally to an increase in communication, which creates a need for knowledge to be translated between scientific and non-scientific fields (transdisciplinary).

This communication is not without issue, and one such problem with communicating scientific knowledge is the expectation that the public have a right to be informed of all scientific developments. Early critics of this problem called the gap between the scientific community and the public a 'gulf of incomprehensibility' (Shapin 1990, 994). This gulf, for them, was proof of the misleading effects scientific discourse can have when extended to the public. By trying to communicate scientific knowledge effectively to a non-scientific audience, scientists will always fall short of the perceived responsibility they have to relate their knowledge to society.

There are further problems caused by this gap, such as legitimacy. Many scientists today feel like they have to legitimise or justify their research, not only to their own community and the wider public, but also to scientists in other fields of scientific research. This leads to many scientists feeling pressured to present their findings in a plausible and simple way to the general public (cf. Hill 2015), and to scientists of other disciplines, in order to legitimise their research (Wilke, Lettkemann, and Knoblauch, 2018; Wilke and Lettkemann 2018).

These issues of communication and legitimacy are deeply connected. In a society where knowledge plays an important role, the institutional pressure placed on scientists to produce knowledge with interdisciplinary or non-scientific relevance has grown in recent years (Hill 2017a). This trend is exemplified by the growing number of science festivals, science magazines, science cafés, citizens science programmes, public science communication events, and transdisciplinary Science Slams. My use of the term science communication in this book refers to the communicative action of scientists. More specifically, when scholars or scientists talk

DOI: 10.4324/9781003172635-1

to each other or to a non-academic using any knowledge they call their 'scientific expertise' (insofar as this reference is essential to the form of communication), I call this science communication. The new genre called the Science Slam 'is a 10-minute oral presentation, in which presenters try to interest their audience, make their talk intelligible and try to establish a slam atmosphere' (interview with Alexander Deppert, 2014). The Science Slam should be understood as an institutionalised genre of communicative action.

The events mentioned above are the result of a recent requirement for communication about knowledge in society. As Knoblauch commented 'the less knowledge is shared, the more needs to be communicated' (Knoblauch 2008a). The rise of a 'Kommunikationsgesellschaft' [which loosely translates into English as 'communication society'] (Knoblauch 2017, 329–377) is defined increasingly by the invention of new forms of communication and communicative genres. As Knoblauch argues, communication culture creates an order which no longer substantiates faith in substantive truth but replaces many written explanations with visual principles. These developments can especially be recognised in transdisciplinary contexts. The growing practices of visual conventions are, to a greater extent, producing legitimacy performatively.

The crisis caused by legitimacy and communication in modern society, where knowledge is of utmost importance, has worsened in the last couple of years. In 2016, 'post-truth'[2] became the International Word of the Year in the *Oxford English Dictionary* after both Brexit and the US election campaign arguably showed that the American and UK public were more influenced by emotional narratives and alternative facts than by boring truths. In addition to showing us how knowledge can be manipulated, the US Election showed how communicating science is a problem, even when it comes to celebrity figures. The challenge that scientists face in trying to get the public to believe certain scientific truths is just as difficult as the challenge many experts faced when having to inform Donald Trump, the former American President, about the intricacies of political relations and world events. While scientists face highly complex issues, advisors to the then president had to deal with a man who has an attention span of about 30 seconds.[3] Trump's advisors, knowledgeable about politics and the American Government and aware of the historical relevance and significance of this knowledge, had to present to a man who only wanted to be entertained by brief descriptions, impressive graphics and in finding out how this information would benefit him. At the centre of the Western World, there was a powerful white man in charge—the President of the United States—who denied certain scientific truths and was uninterested in many crucial realms of knowledge. What scientific texts refer to as 'the man in the street' (Berger and Luckmann 1967), has been replaced today with the 'President of the United States'. Either way, the techniques used by advisors to communicate effectively with the President, and the methods scientists use to communicate effectively with the public, will have a great impact on history. As Hannah Arendts (2017) studies on totalitarianism show, very simple things—like an unwillingness to think, irresponsibility, and a lack of empathy—can become a breeding ground for extreme and evil systems. One could say that the developments in US Politics since November 2016 have much more in common with the Science Slam than one might first think. In both cases, the central goal seems to be to replace boredom with entertainment.

However, this book does not seek to equate Science Slams with interactions with Donald Trump. Instead, this book asks how challenges in communicating and legitimising science in the public are addressed today. I would like to find out how, and why, communicative actions are developed, before turning to assess the uncertain relationship between science and the public. Finally, I wish to evaluate what we can legitimately call science in this context. I will suggest that new approaches of communicating science in lecture-based events can be seen as an alternative way of communicating and legitimising science because they are an attempt to integrate science into the lives of the public. Arguably, the new approaches of communicating science offer an opportunity for the observation of contemporary communication and legitimisation practices. I will discuss to what extent these challenges are reflected in Science Slams, before considering what solutions Science Slams may offer. I wish to answer the question: 'how can scientists today present their scientific findings to a non-academic audience, and how can they make this knowledge relevant to this audience?' The contextualisation of science in a post-truth society, I argue, should draw attention to important issues, and to sensitive topics involved in communication today.

As I have already established, entertainment and the avoidance of boredom seem to be important in modern society. As experiences and communication become more central, the staging of the self becomes increasingly important (Soeffner 2001). In this world, the success of an individual's Science Slam can decide whether an individual receives financial support for their post-doc.[4] *It is important, therefore, to ask critically what consequences this has for the members of society.* While the intellectual public sphere was traditionally led by the universal intellectual, in late modernity he is increasingly being replaced by experts (cf. Pfadenhauer 2010; Eyal and Buchholz 2010). I am assuming that in a time where there are fewer 'universal' thinkers and more experts in specialised fields of knowledge (differentiation), authors of knowledge become more important in public science communication. Even though economic support for scientists is mostly independent from science that is shared with the general public, there is a need for visibility in order to generate trust. Scientists are, therefore, required to share their scientific work with the public in order to stay visible and so the public is satisfied that the scientist is an authority on their subject. It is, in other words, a 'regime of visibility' (Bucher 2012, 1165). As part of their professional duties, scientists are required to produce novel ideas or ways of presenting their research when presenting their work in public (Schnettler and Knoblauch 2007, 270). New forms of communication might have consequences for knowledge itself. This is perhaps most evident in translation processes, in which a shift from text-based communication to visual-based communication may take place.[5]

Obviously, communicating knowledge to an audience requires high standards. Both in and outside of academia, people question how visual-based communication may change or limit the contents or specific interpretation that the author wishes this knowledge to convey. Communication in this way is often assumed to be in service of some kind of external logic, and so aesthetic forms of knowledge are seen as having a greater purpose. Thus, knowledge communication increasingly relies on new information, new communication technology, and new media, the likes of which are continually being developed. New developments and events in public science communication can be understood as a manifestation of society's dedication to the principle of efficiency through a focus on the aesthetics of communication (Schnettler and Knoblauch 2007, 271).

Sociological Perspectives on Science and Its Worldly Demands

This new representation of science is in direct opposition to the classical view of the responsibilities that scientists should have. In the early 20th century, the sociologist Max Weber claimed that scientists should be characterised mainly by their devotion to serve science (Weber [1919] 1992, 15). Weber wanted scientists to have deep passion for their research and be isolated from other societal influences.[6] He claimed that there was a kind of academic asceticism in which bodily excesses, experience, and the focus on one's own personality were condemned. Weber also wanted to distance the labour of scientists' work from non-scientific activities and duties (for example, binge drinking or political and economic activities). Weber's arguments raised serious questions about scientists' relationships with the public. Despite acknowledging the role that inspiration and creativity play in scientific work, Weber firmly believed that scientific work should occur at a distance, away from practical, worldly pleasures and concerns. The state that Weber described as 'dignified loneliness in the service of science', is often criticised today and viewed as snobby, self-involved behaviour, far removed from reality. Critical terms like 'science in an ivory tower' or 'science ghetto' (Rössner 1992, 7) demonstrate a strong opposition to the Weberian ideal.

> The large institutes of medicine or natural science are 'state capitalist' enterprises, which cannot be managed without very considerable funds. Here we encounter the same condition that is found wherever capitalist enterprise comes into operation: the 'separation of the worker from his means of production' . . . . As with all capitalist and at the same time bureaucratized enterprises, there are indubitable advantages in all this. But the 'spirit' that rules in these affairs is different from the historical atmosphere of the German university. An extraordinarily wide gulf, externally and internally, exists between the chief of these large, capitalist, university enterprises and the usual full professor of the old style. This contrast also holds for the inner attitude, a matter that I shall not go into here. Inwardly as well as externally, the old university constitution has become fictitious.
>
> (Weber 1977, 131)

As we see in this quotation, even in Weber's own time, science had become involved in public life. Weber seems to lament the disappearance of the old, historic atmosphere of German universities, the 'old university constitution has become fictitious'. Much later, Gibbons et al. (1994) described this development as Mode 2 of knowledge production. Yet, with the shift towards transdisciplinary research, science has become more open to outside influences even from those not considered to be 'properly scientific'. As a result of this new way there is a requirement to produce certain types of knowledge, and to also make knowledge more robust, so it can stand up to society. Knowledge has to become more hetero-genic, non-hierarchical, transdisciplinary, and above all, useful for society. Etz-kowitz and Leydesdorff (2000) argued that Mode 2 was actually the starting point of science, before the academic institutionalisation of the 19th century took place. The emergence of the knowledge society was characterised, among other things, by a dominant reference to methodically gained knowledge, which was connected to the ideal of objectivity. According to Daston (2001) the ideal of autonomous neutral science (independent from the cultural context) emerged quite late in the post-war period. Since the mid-19th century, 'mechanical objectivity' has become central in science. Mechanical objectivity aims to eliminate all forms of human intervention in nature; either through the use of machines or through the mecha-nisation of scientific procedures (cf. Daston 2001, 153). Today, scholars accept that we live in a time of entrepreneurial science (Etzkowitz 1998). In the United States, at least, an increasing number of scientists have left their 'ivory towers' to become more involved in work in industrial settings. Etzkowitz argues that these scientists combine both the pursuit of truth and the pursuit of profit. In addition to this, the need for financial support outside academia has grown, and universities frequently compete against each other for funding. In this transdisciplinary area, the utilisation of scientific research for commercial gain becomes more likely. As Etzkowitz states,

> Entrepreneurial scientists' research is typically at the frontiers of science and leads to theoretical and methodological advance as well as invention of devices.
>
> (Etzkowitz 1998, 826)

With this external focus, an innovative society (Hutter et al. 2011) has become, in a way, a part of the university. Living in an innovative society means that there is an expectation that social science must produce novel ideas and, as a result of this, many scientists today believe that the advancement of knowledge occurs through innovation. Business-like activities have challenged the traditional, monk-like existence of researchers (Etzkowitz 1998), and entrepreneurial scientists are con-stantly moving back and forth between industry and university. As Daston and Galison (2007) point out, this new scientific persona is a hybrid of science and engineering and combines the ethos of the late 20th-century scientist, with the technological focus of the industrial engineer, and the artist ambition of a creative. Nowadays, an 'engineering-scientific' self is the dominant scientific persona. *In the same way that we acknowledge the change of identity in the working world, which Bröckling (2007) describes as the 'entrepreneurial self', we see a change in professional science in the way self-technologies foster productivity and entrepreneurial self-optimisation for scientists.*

The requirement for scientific research in the world has also increased. The general public view scientists as problem solvers, which sometimes leads to the expectation that scientists should get out of their 'ivory towers', do their research in a socially acceptable way, and present their findings to all, ensuring that it is comprehensible. If you ask a scientist today what their research is about and what the public needs their research for, the scientist is expected to know the answer. Furthermore, they are expected to be able to explain their findings to an external audience easily, and the audience is expected to be enthused by their research. Thanks to their new cognitive style, entrepreneurial scientists should not have a problem with this expectation. Yet, performing can be a challenge for many scientists who work in more self-referential fields. Organisers of Science Slams acknowledge that different types of scientists communicate in different ways.

> And scientists from the humanities have not joined the game to compete for funds like the others. And they do not feel that it is their job to communicate in this way to the public . . . this does not hold true for all of them, there are nuances. But it is a fact that when you search for Science Slam presentations you find that they are predominately from the natural sciences and engineering. This is an indication that the consciousness is not yet there, or those from the humanities are not so keen to be on stage.
>
> (MK#45)[7]

There is a risk of losing complex, scientific research or content if a performance has to be both accessible and entertaining, which is problematic. Scientists have to perform in order to maintain their research, yet such performances jeopardise the integrity of their work. Therefore, in order to stay relevant in the modern world, scientists and scientific communication as a genre has to continuously reinvent new ways of communicating. In my work I ask how, and why, particular actions of communication are undertaken to navigate the complex and uncertain relationship between science and the public. I am also interested in exploring what happens to the scientific persona when scientists breach the division between scientific and non-scientific work, as explored earlier in the work of Weber.

Structure of the Book

The next section explains the structure of the book, presents further insight to my theoretical perspectives, and systematically refers to my empirical questions.

Chapter 2 (Public Science Communication: From Old Styles to New) is a brief history lesson. This book concerns a new genre so it seems necessary to start with the past. Therefore, Chapter Two looks at early forms of science communication and representations of science. We then turn to examine the repression of women in science before looking at the development of science after the Second World War. The reader will see how the scientific obligation to communicate with the public has grown since the 1980s. A brief look at recent developments in public science communication will show the background in which the Science Slam has emerged. After that, I will give a small overview of the staging of a contemporary scientist (re-presented self/ scientific persona). I will outline what happens to the scientific persona when scientists try to overcome the typical conventions of scientific representation.

Chapter 3 (Developing a Theoretical Framework in Which to Study Science Communication) outlines my theoretical frame. After an excursion to the social construction of knowledge in Science and Technology Studies (which will be referred to as STS going forward) and the Sociology of Knowledge (which will be referred to as SSK going forward), I will develop a heuristic that makes it possible to study the communicative construction of science communication from the perspective of the sociology of knowledge. After looking at diverse follow-ups to social constructivism (Berger and Luckmann 1967) in sociology and the sociology of science, I will explain how a triadic structure of communicative construction (Knoblauch 2013) might take shape, and then—based on classical social constructivism—I create a triadic perspective on (scientific and all other forms of) knowledge. This triadic perspective on knowledge will outline a method in which to study science communication and the communicative genre of the Science Slam. I reconstruct the base for my research and introduce my definition of science communication. In this chapter, I will also highlight why we need more studies that focus on science communication processes particularly based on communication and performativity. I will outline the central presumptions I make in my work and address why there is a lack of research in STS, especially regarding embodied interactions in public science communication. Based on the helpful work of Goffman, the sociology of knowledge and communicative constructivism, I will outline communicative actions and material forms that help to establish a scientific body of knowledge in public science communication events.

In Chapter 4 (Material and Methods) I will show how the triadic structure is imitated in my own research design and methodology. I will explain my research process and my empirical data base. I used video analysis in order to understand performativity and situated performance on stage and aimed to learn more about the justification (Berger and Luckmann 1967) of actions using qualitative interviews. I used several other methods, for example, basic content analysis of the Science Slam publicity material, which I will expand on in this chapter.

Following this, Chapter 5 (Science Slam as Communicative Innovation) will apply the triadic perspective to innovation studies. In the past few decades, innovation researchers have done much to complement the economic and technical consideration of innovation with a sociological perspective. In addition to the standard categories of invention and diffusion, I will argue that innovation as a sociological category must pass the test of situated and interpersonal success. My focus on subjective knowledge (justification), communicative actions of people (performance), and the objectivated world (socio-material arrangements, body, and language) show that all three levels have an impact on the establishment and institutionalisation of a new genre. After I clarify what may be considered 'new' or 'better' modes, I question the problems of communication and legitimacy. In self-descriptions of the genre there are references to the current problems of science communication. If my argument that a new genre has been established is true, we must find an indication of broader changes in expectations. In this chapter, scientists themselves describe their often-problematic relationship with the public, and how they embody this relationship. Finally, since new genres should always be seen in a historical context, I will show how several other genres have been institutionalised at the same time.

In Chapter 6 (Science Slam as a Genre), I will predominately focus on the social context and on the distinctive character of Science Slams. In this chapter I will describe some features of the outer and inner structure of the Science Slam including interaction orders, communicative roles, setting of frames, and the presentation of scientific research. I will further analyse the interactional organisation of the Science Slam and outline a specific style of science presentation and science representation. My research is concerned with situated performance on stage, in which the acting subject with socially derived knowledge is seen as a possible source for change and the situated realisation of actors is described as a source of innovation. By looking at scientific (re-) presentations, I am interested in real time gatherings and performances in a physical and spatial context. Special attention will be given to the settings of the performances, since social and spatial order is important in attempting to understand social situations. If it transpires that other spatial and temporal specifications, event regulations, and unspoken information define the situation of the lecture, we can conclude that a new genre has been established.

Chapter 7 (Science Slam in Contemporary Society), concludes my book. This chapter connects the Science Slam to contemporary developments within society.

Notes

1   In this book, the word 'science' will be used like in the German tradition, which means it comprises all kinds of systematic knowledge, about all possible objects, and thus comes closest to the Latin *scientia* (cf. Daston 2001, 137). In French and English 'science' refers mainly, sometimes even exclusively, to the natural sciences.

2   This means after the truth was known.

3   'Keep it short—no 30-minute monologue for a 30-second attention span. Do not assume he knows the history of the country or its major points of contention. Compliment him on his Electoral College victory. Contrast him favourably with Barack Obama. Do not get hung up on whatever was said during the campaign. Stay in regular touch. Do not go in with a shopping list but bring some sort of deal he can call a victory'. Source: www.theguardian.com/commentisfree/2017/jun/01/donald-trump-incompetence-white-house-staff-intervention

4   Applicants for a postdoc start-up financing programme at the TU Berlin are instructed as follows: 'Applicants for post-doc promotion please note: in addition to the conventional written application, applicants for post-doctoral promotion have the task of convincing the members of the structural commission as well as the interested TU public

about their research project within a five-minute lecture (Science Slam)'. Source: accessed January 18, 2018, www.forschung.tu-berlin.de/servicebereich/menue/service bereich_forschung/aktuelles_rss_feeds/.

5   An example of this is drawn from my own personal experience. When I spoke to one member of the audience at a Science Slam event, she said she imagined a Science Slam would be like a Poetry Slam, but to her surprise it was very 'image-orientated'.

6   'Inspiration in the field of science by no means plays any greater role, as academic conceit fancies, than it does in the field of mastering problems of practical life by a modern entrepreneur. On the other hand, and this also is often misconstrued, inspiration plays no less a role in science than it does in the realm of art. It is a childish notion to think that a mathematician attains any scientifically valuable results by sitting at his desk with a ruler, calculating machines or other mechanical means. The mathematical imagination of a Weierstrass is naturally quite differently oriented in meaning and result than is the imagination of an artist and differs basically in quality. But the psychological processes do not differ. Both are frenzy (in the sense of Plato's "mania") and "inspiration" ' (Weber 1977, 136).

7   The abbreviation MS#45 stands for interview number 3. The interviews carried out for this project were enumerated chronologically, starting with 42. These will be used throughout the book.

2 Public Science Communication

From Old Styles to New

In Germany, the organisers of Science Slams are critical of the fact that exchanges between scientists and non-scientists are rare. Their critique rests predominately on the fact that many scientists are unwilling to share their knowledge with the general public, preferring instead to remain in their so-called 'ivory towers'. To fully understand the recent developments in the genre of science communication, it is important to look at past instances of social transformation and at the shifting role that scientific knowledge has had in the public eye.

Early Forms of Communicating Science to the General Public

What we consider to be the 'knowledge-based society' has roots in the 17th century, though it is generally accepted that it formally started in the 1970s (Böhme and Stehr 1986). During the 17th century, the order of knowledge was changing and there was also an explosion of scientific communication. Before the 17th century, the societal role of the university scholar was that of a 'court bureaucrat or the recipient of Crown patronage' (Shapin 2006, 186). The activities of such scholars were incredibly intertwined with Christianity, which was the accepted foundation for both knowledge and hierarchical society (ibid.). Studying science at university was, however, usually just a stepping stone in the career of scientific practitioners. Due to the miserable salary and heavy teaching load, being a member of a scientific society 'had no stable significance for the identity of the 17th-century man of science' (ibid.). At this time, court societies were the principal arenas in which science was publicly demonstrated and therefore judged as trustworthy. This period is also referred to as the 'golden age of scientific amateurs' because experimental demonstrations were expected to be both entertaining and astonishing (Weingart 2005). Shapin and Schaffer (1985) demonstrate how this occurred in the example of 17th-century scientist Robert Boyle. Boyle exemplified how experimental methods were used in public to verify hypotheses; by making something visible to an audience in an experiment, Boyle generated trust for his inventions and findings. While he backed up his discoveries by research in his laboratory, visual and material techniques were used to encourage the audience.

DOI: 10.4324/9781003172635-2

An important development in the sharing of scientific knowledge with the public was through the development of coffeehouses in England. As Johns (2006) demonstrates, in the mid-1650s the coffeehouse was the new place for proposing, debating, and resolving ideas that were already established in Continental Europe. Coffeehouses made new forms of socialising possible, and the very interior of the coffeehouse allowed for this. The provision of a number of tables allowed intellectuals to discuss a variety of topics, with many different people.

The history of popularising science can be linked to the rise and fall of the bourgeois class. The emergence of this new movement in the 18th century introduced new forms of debate, which tested cultural differences. Habermas' account, 'The Structural Transformation of the Public Sphere: An Inquiry into a Category of Bourgeois Society' (Habermas 1990), describes how the bourgeois as a class emerged through developments in French, British, and German society in the 18th and 19th centuries. This change can be seen most clearly in the rising number of those who could read. Whereas before this time reading was something for scholars alone, the establishment of the age of mass communication (which emerged from the invention of the printing press in the 15th century), suddenly meant that a wide and varied audience could consume literature. According to Habermas, this new change in literacy levels led to small groups of townspeople and bourgeois wanting to explore new avenues of public communication (Habermas 1990, 13). British coffeehouses and French salons were new areas that the public could use for 'raisonnements' and for the consumption of cultural knowledge. Habermas elaborates on this idea further through his explanation of the asymmetrical ideas the 18th-century bourgeois had about culture: that the crowd was educated to be part of a sophisticated culture, but that culture was not to be reduced to be a culture of the masses (ibid., 254).

The idea of educating the uneducated is hierarchically structured. It was in this context, one of a sophisticated public bourgeois, that the idea of connecting science to the public in a non-academic environment began to arise. At the same time as this, secular belief in ideas and theories from, for example, the natural sciences was spreading. Subjects that had been dominant for centuries, such as Theology, Medicine, and Law, were beginning to be complemented by subjects like Chemistry and other laboratory sciences. Discoveries such as Copernicus' theories on heliocentrism and Isaac Newton's understanding of gravity, were a catalyst for these changes (Ebel and Lührs 1988, 15). Yet, the real pioneers of these ideas were Francis Bacon (1561–1626) and Rene Descartes (1596–1650). As Sheffield (2004) explains:

> The new men of science had to learn the new scientific methods introduced by Bacon and Descartes and subsequently had to protect these ideas from thinkers who preferred traditional ways of seeing and knowing the natural world.

The historian Daum (1998) describes how these new thinkers came into conflict with their old counterparts during a dispute in which the two sides disagreed about the relevance that teaching natural science had in German schools during the 19th century (ibid., 51). This dispute is a good example of many such disagreements between the upholders of the traditional educational ideal and the new rationalists. The dominant humanist ideal, which held philological leanings, was contested by realists, who aimed to push the relevance of natural science forward. Criticisms of the realists were based on claims that they were utilitarian. Indeed, the ideas of natural science were described in many such ways: as business or utilitarian thinking, as materialism, or as the commercialisation of society (ibid., 53). In contrast to modern views of such descriptions, in the 19th century terms like 'purpose', 'utilitarianism', and 'practicability' had negative connotations. Realists ended up having to argue back against the denunciation of the utilitarian school by using a humanist rhetoric.[1]

In a time of radical change, particularly that of revolution and the foundation of the German Reich (in 1871), it was not just ideas of democracy and natural science that were changing. What also emerged was a new bourgeois concept of the self. Even though the bourgeois mainly followed a philological-philosophical ideal, the notion that humans were capable of understanding the laws of nature was a challenge to the traditional theologically based view. Since, hypothetically, every member of society was believed to have the faculty of reason, the idea that everyone and anyone could have dominion over nature did emerge (Ebel and Lührs 1988, 15). From then on, both the popularisation of science and the education of the public became an ambition of the bourgeois (Daum 1998, 4).

For this reason, a new era known as 'the pioneer's period of popular science' began in 1848 (ibid.). The impact that the associations and clubs that emerged from this had on the transformation of the public sphere and collectivisation of the bourgeois was enormous.[2] While the scientific academies contained groups of elitist scholars and excluded the broader public, the bourgeois reacted to this by creating non-academic spaces in which to debate science. After 1848, public collections of natural things such as chambers of curiosities, museums of natural science, zoos, botanical gardens, observatories, and aquariums were established (ibid., 5). It was a period of public lectures and learning. At this time, academic scholars had also started to favour a new educational ideal which was opened up to the broader public.[3] In addition to debates in clubs and meeting houses, there were wandering orators (*Wanderredner*) who travelled through Germany with small figures and experimental devices which they used to demonstrate scientific theories. In the same period, perhaps not unexpectedly, the media coverage of popular science increased.

For Habermas, the fall of the public bourgeois took place in the late 19th century following the departure from a hierarchy of social classes to a much more heterogeneous society with fewer class barriers. One consequence of this development was the destruction of the economic barriers which had previously stood in the way of education (most importantly, books became cheaper). At the same time, a psychological barrier was maintained. The subsequent triumph of cultural industries introduced a second chapter in the commercialisation of culture.

However, this time not only did the financial barrier disappear, but also the psychological. Habermas describes how, in book clubs after the Second World War, the quality of the content of books was lowered so they could reach a larger audience. Building on Adorno and Horkheimer's criticisms about the cultural industry, Habermas describes how the change from 'acculturating'[4] to a culture of consumption developed. In the emerging pseudo-public area of mass media publication, meetings to discuss cultural topics were replaced by a public consumption of mass media. Habermas does not describe the ensuing mass media critically, but rather considers them as staged, one-sided and undemocratic.[5] If we are to believe Habermas, the source of an educated, intellectual social class in a public sphere had run dry. According to him, the 20th century was a period of public incapacitation and marked by a dissolution of non-academic audiences from the academic world. After the First World War, in the scientific community it was generally believed that non-academic people lacked an understanding of scientific principles.

Habermas is often criticised for his explanation of the past golden age of public 'raisonnements'. Nancy Fraser (1992) argued that Habermas' concept of a 'bourgeois public sphere' falsely portrayed the idea of a single public sphere and in doing so, idealised it. Fraser suggests that Habermas supports the illusion of a free and equal public discourse with open access. However, Habermas never mentions the fact that his ideas are built on the notion that differences, namely the exclusion of gender, class, and the alternative public, are unimportant. Fraser, however, notes that the public bourgeois was not only an idealised utopia; it was also a male-dominated, ideological concept that only supported and legitimised the leadership of men. Therefore, Fraser argues, Habermas' claim that those involved in disputes could argue as if they were equal, should be replaced by a recognition that women and other non-dominant social groups were not allowed to participate, so any notions of equality should be dismissed. Fraser presents an alternative idea of what post-bourgeois publics could be: subaltern counterpublics are 'an exemplary form of parallel space which through its activities acted as a bulwark against patriarchal hegemony'. For Fraser, the formation of identity is part of public participation:

> Public spheres are not only arenas for the formation of discursive opinion; in addition, they are arenas for the formation and enactment of social identities. This means that participation is not simply a matter of being able to state propositional contents that are neutral with respect to form of expression. Rather . . . participation means being able to speak in one's own voice, and thereby simultaneously to construct and express one's own cultural identity through idiom and style.
>
> (Fraser 1992, 166)

The works of Hannah Arendt (1960, 1970a) may be of interest here. Arendt compared the public sphere to the ancient Greek polis, a place where political action was built in the centre of the city where people met to debate ideas. In the polis, people were confronted by others who held differing opinions to their own, which was indication of the plurality of society. In this context, there was a need for people to position themselves in relation to others; indeed, it was a basic condition for the political. According to Arendt, therefore, power is not a one-sided domination by one political figure, but something that arises between 'the people' when they act together, and the world is created and shaped in this joint action between people. For Arendt, the public space is not just there, it emerges through the power of action, and through shared opinions. Politics, meanwhile, emerges through the communication between people.[6]

I understand public spheres as communicatively constructed spaces in which different perspectives and identities relate to, and connect with, each other (cf. Fraser 1992; Arendt 1970a). By extending deliberative models (cf. Habermas 1990), more recent studies understand the public sphere as an empirical category. The public sphere is neither normatively postulated nor ontologically presupposed, but communicatively constructed through public communication and media-based procedures. I empirically relate situated public spheres between embodied subjects, objectivations, and others. I note that public spheres are produced in performative acts and wish to analyse how the boundary between science and the public is enacted and performed in Science Slam presentations. My project further aims to answer the question of how publics are communicatively constructed (Knoblauch 1995). In this book, I will refer to the public in a post-bourgeois manner. Additionally, I will argue against the narrative of cultural decay.

Next, I will give an overview of the historical background of the scientific self. This is very important as it allows us to compare the scientific self in the past with the scientist who is involved in Science Slams today. As I have briefly noted, modern views of scientists are often challenging. In fact, scientists have been labelled as problematic throughout the last 200 years. To fully understand this problem, I will present a few historical studies that highlight this issue, all of which are primarily based in academia.

Early Representations of the Scientist

> The image of the white male scientist in his pristine lab coat, intently working away in his laboratory, is the current icon of the scientist, but this perception may change.
>
> (Sheffield 2004)

This description is one that is commonly associated with scientists. This may seem like a modern idea, however, as early as the 18th century, the image of the scientist became the focus of attention in portrait paintings. In these paintings, the scientist was portrayed in a number of ways: as carrying out theoretical work at their desk; as an experimenter; as a proud inventor surrounded by colleagues or students; or finally, as a lecturer in lecture halls (Krifka 2000a). Although these themes varied, the scientist was usually surrounded by books, inventions, models, or his scientific peers. Being portrayed in such a way was mostly positive. Such ideas about scientists came about primarily during the French Revolution and the Scientific Revolution, both of which went hand in hand with influencing the public's perception of scientists. With the arrival of the Enlightenment, empirical, natural science became dominant, and the institutional sphere of knowledge changed. The teaching and sheer sanctimony of traditional Christian beliefs were challenged and replaced by empirical sciences that focused on discovering new knowledge. The transition of curiosity (*curiositas*) from a mortal sin (Augustine) to a venial sin (St Thomas of Aquinas), and later to a virtuous characteristic (Thomas Hobbes) made the search for the mysteries of nature more likely (Krifka 2000a). Virtues such as working hard, cultivating an eagerness to learn, taking great pains over knowledge production, and being accurate became characteristics of a scientific person.[7] The religiously favoured ideal of the polymath was replaced by the ideas of the experimental natural scientist.

This does not mean that natural scientists were without challenges or problems, however. For a long time, dominant social groups were suspicious of intellectual specialisation and scholarly isolation (Shapin 2006, 190). The stereotypical image of the scientist suffering from social misery is not a new one and can be widely found in documentation from as early as the medieval period. Negative connotations ascribed to the scientific persona have been present throughout the last 500 years or so, and descriptions vary from the socially isolated, crazy alchemist to the mad scientist (Neumann and Fuchs 2009; Summer 2008). Indeed, scientists were often portrayed as being so engaged with science that they forgot all secular affairs (cf. Daston 2003).[8] To be involved with science was often described as an obsession with the subject, which also involved isolation from society. Just as the sacred principles of classical academia had involved financial and personal sacrifices, scientific work and domestic life were also difficult to harmonise. The major concerns about scientists, as noted in various pieces of literature, were that scientists were: naïve, egoistic, reckless, obsessed, crazy, narrow-minded, unethical, without restraint, wasteful, arrogant, and finally, not able to see that their research might have unforeseen consequences and may lead to a disaster (Neumann and Fuchs 2009).[9] There was also a fear of scientific nihilism, atheism, and materialism, as well as the unthinkable idea that in his delusions of grandeur, the scientist would see himself as above the Divine.

As suggested above, the academic rigours of scientific discovery were at odds with domestic life. By the 19th century, however, Daston argues that the scientific persona had begun to become domesticated. According to Daston, scientists in the 18th century, such as Charles Darwin, feared becoming 'workers' at the expense of family life and thus married to prevent a life solely filled with work. Despite this argument, the compatibility of a social life or a successful family life with science was still a challenge. To be a scientist at this time was to 'transform the self', not only in terms of values and emotions, but also in terms of physicality (Daston 2003, 113). Historically, there have been many different ideas as to how one can objectively be a scientist (Daston and Galison 2007). In earlier periods, the ideal type of scientist was the wise scholar, who standardised or idealised nature's variability. In the last 200 years, the ideal type of scientist was the hard worker.

> It was no longer variable nature or wayward artist but the scientific self that posed the greatest perceived epistemological danger.
>
> (ibid., 198)

As Daston and Galison show, in the 19th century, self-elimination became an imperative. In this era, subjectivity became the enemy in the goal of eliminating the human element. Feelings and emotions were to be set aside, while self-control and scientific discipline were favoured. If the data were mechanically mass-produced, this demonstrated authenticity.

The societal pressures placed on scientists required them to orientate their self-concept in the direction of the most powerful group in society. Correspondingly, Shapin (2006) argues that the man of science in modern times has changed his demeanour from that of an isolated, unfriendly, confrontational, authoritarian, individualistic, dry, and pedantic character to one with a more gentlemanly manner.[10] In the culture of the 17th century, the aim was to be civilised, masculine, rational, thinking, distanced, ascetic, and text-oriented as opposed to body-oriented.[11] The new intellectual order established in the 17th century made it acceptable to create knowledge in public (Shapin and Schaffer 1985). I have already mentioned Habermas' (1990) argument that the bourgeois class emerged in the context of developments in France, Britain, and Germany in the 18th and 19th centuries. Although Habermas describes how the bourgeois came about, Mergel's description of the bourgeois way of living is key. Bourgeois culture can be described as a way of living in which the commitment to a community spirit, the need to construct individuality through socialising, the participation in a masculine culture, and practices of self-monitoring are central (Mergel 2001).

We now turn to look at the position of engineering and engineers within the bourgeois culture. In Germany, engineers tended to have underprivileged beginnings in the 18th and 19th centuries. Historical research on the bourgeois class in Germany during the 19th century indicates that engineers had to deal with a lower status than other middle-class intellectuals.[12] Jarausch (1990) shows how many engineers attempted to reach the same status as other professionals and how they often had to fight against the stereotype cast upon them as an uneducated and odd social group. Jarausch further argues that, in compensation, engineers preached a technical ethos of progression and argued that the improvement of welfare could be reached through material innovation which they could provide.

As Neumann and Fuchs (2009) show, the idea of the mad scientist in 19th-century literature was so popular because of the discomfort many people felt regarding the way science was at odds with traditional values. There was also a general misunderstanding of new scientific ideas among the general population, which only enhanced the discomfort. Conservatives and modernists (even those who were committed to humanistic ideas) equally feared fragmentation, specialisation of knowledge, and the loss of orientation and order which was associated with scientific discovery at this time. It was not so much the fear about scientific and technological advancements, but more a concern about how the fragmentation and specialisation of knowledge would disadvantage traditional universal scholarship. It was the development of experimental laboratory sciences (predominately chemistry) at the expense of humanistic science, which caused these fears.

The relationship between science and society in financial terms changed at this time too. Being a scientist had historically been a job for rich amateurs (Bacon, Darwin, Humboldt), but this evolved to become something that was done by professionals and the technically specialised. Patrons who had supported science with their private funds, such as Alexander von Humboldt who spent his inheritance on scientific research, were replaced by professionals who came from various social standings. This development, the professionalisation of the scientist, created an environment in which the social background of a person was not as important as it had been in scientific communities.

The Repression of Women

As Sheffield (2004) shows, the origin of the view that scientific work was predominantly a male activity began in the 17th century. One consequence of this was that the image of the 'Goddess of Science' completely disappeared by the 18th century.[13] This change, from domestic to institutional science gave women few chances to participate.

> The best they could hope for was to be silent and invisible assistants, a far less powerful position within the scientific community, although not necessarily a less important one for the work of science.
>
> (Sheffield 2004)

Although admittedly some institutions accepted female scientists, women mostly had to deal with exclusion from the official protocol (Findlen 1999).[14] Science studies have shown that science in the last 200 years or so was traditionally oriented towards a patriarchal society. This was not to last, and in the 19th century, the women's movement celebrated success at several universities in Germany. Female demands for a right to an education were met and became a key area of reform. This started in Switzerland, then moved to Germany and by 1900 women were accepted to study at many universities. This change was largely driven by the bourgeois women's movement.

The progress made at the start of the 20th century was not to last. The seizure of power by the National Socialist German Workers' Party (NSDAP) removed many of the laws that the bourgeois women's movement had established. For Hitler, there was only one role for women in the National Socialist Party and this was to reproduce. The NSDAP revitalised ideologies of motherhood and reversed many legal rights that had been achieved. Women were pushed out of public service, lost the right to vote, and their access to university was reduced to a 10% quota. In addition to this, women were no longer allowed to be admitted to habilitation. The

Nazi Regime also succeeded in delegitimising many female, Jewish, and politically opposed scientists, and banning many fields of scientific expertise (the SS took over many fields of law). The Nazis decided that science was to be the ground worker of the regime, and a new type of ideological scientist was established.[15]

After the Second World War, opportunities for female scientists multiplied and thousands of women became scientists and engineers. These female scientists often took the following advice: 'that the only way for women to succeed in science is to practise science like men, only better, working harder, and denying themselves a life outside of their career' (Sheffield 2004, 184). Yet, the problem of women's invisibility continued, (ibid.) and the idea that science was primarily a male activity perpetuated. Despite successes and the increase in the number of women in science in the 20th century, 'the image of the male-only scientist has persisted' (ibid.). Only a few iconic female scientists from the past (such as Marie Curie)[16] are viewed today as successful.

The Scientific Persona After the Second World War

It was not just the relationship between women and science that changed after the Second World War. In society in general, there was a nostalgia for the bourgeois ideals of the past, after 1945 (Hettling and Ulrich 2005).[17] After the collapse of civilisation during the Nazi Regime, academics in Germany had to rebuild their self-image and regain their economic position. This proved successful and in the 1950s, scientists in Germany managed to re-professionalise (Jarausch 1990). Independence became an important value for professional scientists (Gibbons et al. (1994) called this 'Mode 1').[18] As Jarausch points out, many scholars publicly distanced themselves from the Nazi Regime, but the majority of academics hid away in their ivory towers. This meant, therefore, that scientific demonstrations to the public declined (Smith 2009) and the objective expert became the dominant scientific persona (Daston and Galison 2007). The recognition of classifications and patterns became a virtue.

The rise of the inventor or entrepreneurial type (Schumpeter 2000; Etzkowitz 1998) in the scientific community must be viewed in the historical context of the 20th century, when the focus changed from being dogma-based to a search for new knowledge. With this change, a new, experience-based method of researching was established. Breaking up the old and creating new traditions became attractive. This also made the philosophies of utilitarianism more likely to be trusted.[19] Scholars today argue that we live in a time of entrepreneurial science (Etzkowitz 1998). Entrepreneurial scientists are constantly moving back and forth between industry and university, combining the pursuit of truth with the making of profit. Innovation has become part of the university because many scientists believe that the advancement of knowledge is the same as innovative research. Business-like activities have challenged the monk-like existence of researchers (ibid.). Nowadays, an engineering-scientific self is the dominant scientific persona. This is not unprecedented, since by the 17th century the ability to demonstrate scientific principles had already become part of the scientists' repertoire. Science no longer consists of thoughts based on religious doctrine and of the public witness of scientific experiments, but of business-like thinking. In society today, the need for communication has increased and images play a big role in knowledge production and communication. Images have almost become part of the scientific apparatus. In the case of nanotechnology, for example, seeing and doing have merged together (Daston and Galison 2007). The scientist of the 21st century combines the ethos of the late 20th-century scientist, the mechanical orientation of the industrial engineer, and the aesthetic of the artist.

> The image-as-tool seems to enter the scene inseparably from the creation of a new kind of scientific self- a hybrid figure, who very often works toward scientific goals, but with an attitude to the work that borrows a great deal from engineering, industrial application, and even artistic-aesthetic ambition.
>
> (ibid, 413 [sic])

In scientific movies, journalism, literature, and other media in the 21st century, we find a new image of the scientist, although the dominant figure is still the elderly man with dishevelled hair, who wears a lab coat and glasses and who works, obsessed and lonely, in his laboratory, tinkering with a major invention with which he wants to change the whole world (Neumann and Fuchs 2009).

Communication Since the 1980s

In the late 20th century, there emerged a model in Europe that confronted scientists with their social responsibilities. In the 1980s, the British Royal Society warned the scientific community in its programme, the Public Understanding of Science (which shall be referred to as PUS from here on), that science should communicate more with the public and thus inform general society about scientific advancement. The paper urged scientists to improve scientific education in schools, improve cooperation with the scientific committee, and increase their communication with the public.

> Scientists must learn to communicate with the public, be willing to do so, and indeed consider it their duty to do so. All scientists need, therefore, to learn about media and their constraints and learn how to explain science simply, without jargon and without being condescending. Each sector of the scientific community should consider, for example, providing training on communication and greater understanding of the media, arranging non-specialist lectures and demonstrations, organising scientific competition for younger people, providing briefings for journalists and generally by improving their public relations.[20]

PUS was established in the 1980s because science was suffering from a lack of public support, and the Royal Society believed that this was due to the general public's lack of scientific knowledge.[21] With this worry in mind, the Royal Society created several programmes which aimed to improve this lack of knowledge (for this reason their approach is called a deficit model). PUS intended to increase the public's understanding of scientific knowledge.

PUS's central notion, of communicating higher forms of knowledge, was largely built on Plato's ideas on enlightenment, as demonstrated in his famous Cave Allegory. In the late 20th century, ideals of human perfection and issues in developing society were seen as connected to a demanding act of liberation through communication between an expert and a general member of society. Consequently, since the 1980s there has been a general European model set in place for popularising science, which confronts scientists with their responsibilities. STS, who criticised this political programme (Wynne 1992), focused instead on 'policy-informing forms of engagement and dialogue' (Davies et al. 2009). Jasanoff (1997) questioned how one can design a dialogue that leaves space for input from general society and avoids civic dislocation. Other scholars wondered whether the deficit model is really dead, or whether it coexists with other communicative designs (Bucchi and Trench 2008).[22] In reaction to critiques from STS, the programme called Public Engagement with Science (which shall be referred to as PEST from here on) emerged, which aimed to democratise the former approach and create a dialogue between science and the public, instead of indoctrinating and instructing it. As it turns out, PEST proved to be a force for good in science, because it allowed scientific resources to be acquired and provided. STS assessed any improvements as a result of this approach by creating a dialogue-setting (with a special kind of agency).

> In these discussions, "public engagement" has tended to mean deliberative public participation in science policy, rather than the informal education or leisure activities that characterize science communication.
>
> (Davies 2014, 1)

The public communication of scientific knowledge that does not seek to directly inform policy is relatively understudied (cf. Davies et al. 2009). STS's research on science communication has been somewhat immaterial in the past (Davies 2013). There are huge research gaps within STS when it comes to things like materialism, aesthetics, and emotion in science communication:

> Science communication is focused on the elicitation of emotions such as enthusiasm, interest, outrage, and delight, and it often mobilises affective, aesthetic, or material configurations or techniques in doing this. It emphasises that public interactions with science are always grounded in material realities, are always emotional, and always go beyond discursive exchange or arguments. Science communication may be messy, loud, immersive, or reflective.
>
> (Davies 2014, 2)

Public Science Communication Today

In recent times, the reduced number of universal truths and the vastly differing views on how we gain knowledge is challenging what role experts play in communicating to the public. Today, we find a growing number of experts in various specialised fields of knowledge, instead of the traditional universal thinker, although many experts are quite heterogeneous. This diversity of knowledge creates problems too, mainly to do with legitimacy. The problem of communicating scientific knowledge in this era of mass communication is that the public gains access to science mainly through the media.

We can observe that new genres of science often borrow ideas from genres of the past. Lately, more and more scientific events aimed at the public have been developed, which often try to reinvent science in popular and artistic ways. In events such as 'Science Slams' and 'Lecture Performances', scientists are asked to present their knowledge in new ways and to speak to an audience of the public in order to legitimise and communicate science. Some of the events claim to have huge audiences, despite their complex scientific topics. Lecturers aim to connect to the public by means of emotional attachment and artistic demonstration. By relating themselves to different disciplines and genres (like performance art, theatre, and the Poetry Slam), they have developed new strategies for communicating science. The Science Slam, among other events, addresses a diverse audience and uses stylistic strategies taken from Poetry Slams in order to present scientific findings. Art-related events, like the Lecture Performance, are at the interface between instruction and entertainment. There are recent instances of new scientific public events being established across the world. The rise of science festivals (Bultitude, McDonald, and Custead 2011), science magazines (Born 2015), science cafés (Dijkstra and Critchley 2014), citizen science programmes (Irwin 2001), 'long nights of science', and science museums suggest that the Science Slam (Stimm 2011) might be part of a new environment of communication in science. Innovations within the science communication genre have been occurring since the 1980s. Events like TED Talks (founded in 1984), Café Philosophic (founded in 1992), and Café Scientific (founded in 1998) can arguably be viewed as the catalyst of a new communicative movement. This can be seen today across the world: for instance, in the United Kingdom there are events called Science Showoff, Bright Club, and FameLab; in the United States of America they have Science in the Pub, Smart Night, Secret Science Club, and Entertaining Science Cabaret; Australia has Philosophy in the Pub and Fresh Science. Many different scientific events take place in Germany and in many other countries. The goal of such events is to translate scientific ideas into new settings and to enable communication between non-professionals and experts.[23]

The Transformation of Science Communication

As we have previously explored, Daston and Galison (2007) argue that there is a new scientific persona in the present day, which hybridises science and

Table 2.1 Different Types of Scientific Persona and Their Characteristics

Scientific Persona	Characteristics of Successful Science Representations in Science Slam
Human Man	The 'human man' presents the scientist as a human being. The scientist humanises himself by communicating and builds an emotional bond with his research object.
Popular Culture Man	The 'popular culture man' portrays the scientist as an unconventional and cool guy. The scientist is a popularised version of himself. He demonstrates his knowledge of popular culture and translates his activities to fit into the language of the masses.
Working Man	The 'working man' depicts the scientist as a hard-working labourer. In this persona, the scientist brags that he works a lot, and that science is a passionate, fulltime job.
Non- Pseudo Man	The 'non-pseudo man' argues that he is the real embodiment of science (unlike media representations).
Modest Man	The 'modest man' represents the scientist as loser-chic. He comes across modest and gives the impression that his worldly desires are small.
Entrepreneurial Man	The 'entrepreneurial man' presents himself as an inventor. He argues that his inventions are useful in everyday life to regular members of society. For him, the Science Slam is an additional way of finding financial aid for his projects.
Engineering Man	The 'engineering man' portrays himself as an inventive scientist. He argues that he is able to improve technology and is perhaps even capable of saving the world.
Success-Seeking Man	The 'success-seeking man' aims only to win the competition and to impress women.
Ascendency Man	The 'ascendency man' argues that the scientist is superior. This is due to the fact that scientists have better access to knowledge and images than the public. Everyday knowledge amuses him and is beneath him.

engineering to create a type of engineering-scientific self. In the introduction to this book, I mentioned that people are often required to present the outcomes of their work in public as part of a viewed obligation to communicate (Schnettler and Knoblauch 2007, 270). Bearing this in mind, when looking at a model of what a scientist should look like today, one should not expect an obsessed or bourgeois scientist, but rather, either an entrepreneurial scientist who seeks to create new concepts, or an engineer-type scientist who wants to improve things. In my research, I found evidence which proves this perspective.[24] As part of Science Slam performances, speakers often represent different types of the scientific persona.[25] In my empirical work, I discovered that slammers typify themselves. Some types of scientific personas that I witnessed in Science Slams can be seen as an indication that new science communication is growing.[26] I will use, as examples, some successful slams to characterise various types of scientific persona and to emphasise that one scientist often has multiple personas. This perspective makes comparing the represented self in Science Slam performances possible.

This overview shows new forms of popular scientific personas and, as we will see later, many of these are present in Science Slams. The gentleman from early modern times, who was trusted to speak the truth, has been replaced by new types of scientists in a still emerging, rich field.

There are several strategies that slammers use in order to establish themselves as legitimate scientific speakers. The 'human man' type of scientist tends to humanise himself and to build an emotional relationship with his research object. This type of persona probably responds to the historically based fear of social disconnection. The 'popular culture man' exhibits a popularised self, wherein the scientific gentleman merges with popular culture. By showing his ability to make multiple references to popular culture and by rewording his vocabulary to fit into popular culture, this type of scientist aims to show he is unorthodox and cool. The next type of persona, the 'working man', speaks in an authentic way. He makes clear that, for him, research is a passionate, full-time activity which offers pleasure and fulfilment. The 'non-pseudo' scientist argues that, as opposed to representations of scientists in the media, he is the real thing. Next, we have the 'entrepreneurial man', who believes that his role of creating inventions is useful in the practical real world. He looks for financial support from his audience to make new inventions possible. A further type, the 'engineering man', presents himself as an inventive and helpful guy. By improving technology, he aims to save the world. The 'success-seeking' man, however, is primarily concerned with impressing the opposite gender and winning. Finally, the 'ascendency man' argues that he, as a scientist, is more knowledgeable than the public. He also believes that mundane knowledge is amusing and beneath him. As varied as these types may be, all of them imagine themselves to be authentic and creative. One thing that can be agreed on, however, is that the typical 'bourgeois' scientist has been replaced by new and diverse types.

Identity Formation and Public Participation

As we have learned, the aim of connecting science with the public is not new. The history of science and its relationship with the public reveals that there has always been a struggle with knowledge between the average man and experts. If we believe Fraser's arguments, science communication must be seen as a specific cultural setting with certain rules that include some, and exclude others. In Fraser's argument, identity formation is a big part of public participation. In this context, the view of social identities in the past, as described by historians like Shapin, must be viewed in a different way. Many feminist academics agree with this and argue that the body of knowledge that the Western male scientist holds is different to that of women. Haraway (1997) points out that the era of the modest witness, with its distinction between the knower and the known, supports established inequality. She suggests that the leading representatives of science at various points in time might be part of a wider context of a historical power struggle. She is mistrustful about the way Western scientists avoid positioning themselves or their responsibilities in regard to their research. She also opposes the 'invisible conspiracy of masculine scientists and philosophers' and the 'embodied others, who are not allowed to have a body, a finite point of view' (Haraway 1988, 575). The notions of disembodiment and universal claims were described as part of the male Western scientist's bag of 'God-tricks' which Haraway sees as a view from nowhere. For this reason, Haraway calls positioning a key practice in opposing both typical visual tricks and the powers of modern scientists who make 'various forms of unlocatable, and so irresponsible, knowledge claims' (ibid., 583). It is important to bear in mind that in a society there are always multiple coexisting public spheres that reflect differences in society. Fraser's ideas about subaltern counterpublics have consequences, not only for ideas about democracy and political equality. It also serves as a reminder that the communication of science is not open to everyone. Societal inequalities prevent certain groups from becoming part of scientific discussions. In short, they exclude certain parts of the population.

Emotions and Trust

Post-truth politics, emotional narratives, and a decrease of trust in scientists has created a problem for science communication. In order to explore this issue further, I will now contextualise science communication within contemporary society.

Classical sociology describes changes that result from modernity as a release from traditional social relations. Classical sociologists tended to argue for a loss of significance with regard to emotions and social trust. According to Weber, for example, a characteristic of community formation in the classic sense was a social relationship where actions were based on the perceived affective or traditional unity of its members. For Weber (1976), the rise of instrumentally rational forms of organisation suppressed affective action in which the human being was pushed into a cold rational society, while losing collective solidarity.[27] Likewise, Tönnies describes modern society as a circle of people that are primarily disconnected to each other and not, as in pre-industrial times, connected in affective closeness. Durkheim (1981) describes the loss of emotional relationships in the evolution of modernisation as a mechanical to organic solidarity. For Elias (1969), a growing control over our emotions was a central force in the process of human civilisation.[28] On the other hand, Simmel (1964), describes how social feelings were a precondition for society to succeed. Simmel also describes how a reification of emotional feelings in the financial sector occurred. The dichotomy between rationalism and emotion is a prevalent factor in classical sociology. While emotions and traditional social relations were associated with the original, the primitive, the community, the physical, and the feminine, rationalisation and the objectification of social relations are a clear characteristic of modernity.

Since the 1970s, emotions and social relationships have been identified in sociology as a feature of modern societies. Prior to an increase of scientific interest in emotion, there was a growing interest in social relationships which has been described as an 'emotionalisation of all spheres of social life in late modernity' (Greco and Stenner 2008, 14). This was not a return to pre-industrial social beliefs, but a cultural re-evaluation of emotional feeling. This renaissance followed the idea that emotions and social relationships are not necessarily dominant over rationalism or morality, but rather rationalism and morality are dependent on emotional feeling. Emotions, which were in a sense forgotten about, suddenly became interesting, even in fields which were deeply connected to rationalism.

A further result of modernisation was that new ways of building communities became important (Hitzler 1998). According to Hitzler, the post-traditional community-formation implied that the commercial re-formation of communities would take place.[29] An essential feature of these post-traditional communities is that membership is a question of free choice (you can cancel it at any time). Members have to be persuaded to participate (ibid.) and membership is based on shared ideas. Meanwhile, Schulze's (1992) description of the experience-based society starts with the assumption that society today works through differences in experience and in understanding aesthetics. He argues that, in contemporary society, players try to maximise their subjective experiences and pleasures.[30] Schulze ends by describing a so-called 'experience economy' in which those who provide experiences compete against each other.

This post-traditional community building happens in the context of cyborg-ish entanglements of bodies and technologies. In the digital society, social action is built to a greater extent between human behavior and technical processes. The significance of technical agency is given varying degrees of importance in the sociological debates on technology (Rammert 1993) and in the debates of STS, for example, the 'Actor Network Theory' (Latour 2008). Technology, in a broad understanding, encompasses not only artefacts but, very generally, the embodiment of utilitarian means. Since Knorr-Cetina (1988) believes that contemporary technologies are both: usable objects, and epistemic things (e.g., the computer), she proposes the advanced category of knowledge object, which includes technologies—in lieu of viewing technology as disparate unit. According to Knorr-Cetina, knowledge objects have a fundamentally open character. Following the work of Mead (1982), she conceptualises human-object relations in a similar way to social relations. In the research processes she analyses, she observes representations or substitutes, which she describes as an 'object deficiency'. In the relationship, she observes a lack of objects and corresponding structures of wanting and willing on the part of the researchers. Scientific objects and technologies have the potential to be unrestrictedly immutable, and thus remain unapproachable and alien. She uses the research activity of the biologist McClintock to show how the relationship between knowledge object and researcher is shaped. By role-taking, researchers can literally 'become the phenomenon' or 'perceive the object within the internal processing environment'. McClintock describes how she, as subject and researcher, participates in the world of objects and how the object worlds participate in her. This is what Knorr- Cetina calls 'crossover'—that is, reciprocal communicative participation, that is, parts of the subject enter or become the object and vice versa.

Kaerlein (2018) argues that portable computers and smartphones have become an integral part of people's wardrobes and everyday life. This relationship is particularly complex, he says, because people no longer use the smartphone merely as an everyday aid, but also associate emotions such as sadness, joy, happiness, or stress with these devices and their programs and apps. In other words, people experience a personal affective connection to these objects. Kaerlein's approach is similar to Beer's (2016), who suggests that media and body are closely entangled. The neoliberal orientation of society not only influences how we view ourselves, but also how we feel. If metric power (ibid.) and measurements in a neoliberal society affect bodies (e.g., by causing uncertainty and stress), then technical devices and bodies are certainly in a close relationship.[31] This impossible demarcation between nature, man, and technology might best be described by Haraway's (1991) concept of the cyborg. Cyborgs are cybernetic organisms, hybrids of machine and organism and creatures of social reality as much as fiction.

Crucially, rationalism and morality are dependent on emotional feeling in politics today. Emotionally based problems have become an issue for science and science communication. Populist parties have become stronger in the United States and across Europe. In post-truth politics, some politicians are interested in influencing the public based on 'alternative facts' and many of them use arguments based on emotional narratives.[32] According to many scientists, the rise of populist parties has become a crisis for both the legitimacy and the communication of science. Certain scientific findings (e.g., climate change) are denied and disbelieved by some of the most powerful people in the Western world. Additionally, the intersubjectivity of scientific knowledge is questioned by many of those in charge.[33] Since November 2016 (when Trump was elected), Latours' (2003) fear that the disbelief of the masses in scientific fact would create an ideology has come true. Until the global pandemic of 2020, science in the public eye was partly represented by children looking at the Fridays For Future campaigns, for example. Its leading figure, Greta Thunberg, called on the public to get involved in scientific discourse to prevent their own demise. In her opinion, people should use the power of democracy to make their voice heard in politics.[34] She is committed to preserving the public's ability to help shape science and society. Interestingly, Thunberg has pointed out that 'in the United States, climate change is treated as a matter of belief, while in Sweden, climate change is true knowledge for people'.[35] This has the capacity to create a problem in the approach towards the concept of knowledge. Indeed, many non-academics now no longer believe in the doctrines of science. Trust in scientists is decreasing.[36] This distrust might have something to do with the rise of social media platforms and the 'disappearance of trusted gatekeepers who used to have the task of quality control' (Weingart and Guenther 2016, 8). The 'alternative facts' that are presented to us often become independent ideas once they are shown on social media channels or private TV stations. This leads to them becoming a kind of alternative reality for some. In addition, some politicians are de-stabilising the institutions and the credibility of science, and therefore social constructivism.[37] In the context of post-truth politics, it seems more important than ever to constantly reference scientific methods, practices, and techniques so that scientific knowledge will still be trusted.

Before I consider what this means for science communication, I will quickly attempt to portray how dangerous these 'thoughtless people' are, and how they can destabilise political institutions and democracy.

Crisis of Democracy

The sociologist Beck (1986) claimed that since the end of the 20th century we have been living in what he called a 'risk society'. Many ecological, social, and other risks have been caused by industrial progress and the state's existing control mechanisms cannot handle these. To give a few examples: we knew that we could die from radiation from atomic bombs when they were created; we know that climate change could bring about our end; it had been researched that microplastics or antibiotic resistance could destroy our organisms; and we had an idea that religious conflicts and the consequences of capitalism would bring us to our knees. However, we have forgotten that many people's unwillingness to think, combined with their irresponsibility and lack of empathy, is one of the most dangerous combinations in modern society.[38] Democracy, and the freedom that comes with it, cannot be taken for granted. Not too long ago, we experienced what happens when knowledge and ideology become confused. People start to doubt scientific procedures and learned knowledge or misuse them to confirm preconceived unconfirmed conspiracy theories. Conspiratorial or superstitious constructions and theories can easily spread via social media and internet platforms, and operators do not stop it because they have an opportunity to financially profit from it (Langley and Leyshon 2016). At the moment, we are witnessing a dangerous interplay of technology and politics, a toxic pairing of ideological longing for the past and a need for technological futurism. Ebner (2019) argues that extremists want to challenge democracy and build radicalisation machines in order to destabilise the legal order of society. Current radicalisation phenomena, such as the rapid rise of the Qanon movement, can be attributed to the building of a community in a digital form. We have just experienced how a politician with ideological views can destabilise the political and legal world order (Donald Trump).

Therefore, contemporary society faces several new risks that some of us could not have imagined a few years ago. We are in the midst of a crisis of representational democracy. In her book on totalitarianism, Arendt (2017) argued that totalitarian regimes emerge when the feeling of needlessness becomes dominant for many people in mass society. When democratic parties can no longer interest an increasingly apathetic mass, the mass can be drawn in to populist movements. Similar to Habermas, Arendt claimed that the fall of a class society is a problem for representational democracy and political parties, because it represents the loss of class interests as an integrating function. According to Arendt, in a mass society the alienation of political elites from their democratic base is guaranteed.

Sadly, current political events indicate that she is right. On January 6th 2021 a group of Donald Trump's supporters stormed the Capitol in Washington, D.C. This event was a violent attack on the Congress of the United States, and on democracy. The rioters aimed to prevent the result of the 2020 presidential election from being confirmed, as Donald Trump had not won. Instead of accepting defeat to Joe Biden, Trump widely contested the election results via his social media pages and told his supporters that electoral manipulation had caused him to lose the election. The narrative of election fraud spread quickly in traditional mass media.[39] As a result, in December 2020, polls showed that many citizens believed that the election was or would be rigged.

The purpose of this book is not to explore Arendt's theories in full, but a reference to Arendt is helpful in expanding on sociological theoretical approaches with political theory. When misinformation spreads through social and traditional mass media it can lead to a dangerous situation, at which point we should remember Arendt's critical reflections on politics and totalitarianism (ibid.).[40] The people who met to storm the Capitol should not be compared to the ancient Greeks, who met in the polis. Instead, they must be described as violent, 'power and violence are opposites: where the one rules absolutely, the other does not exist' (cf. Arendt 1970a, 57).[41] If we are to apply Arendt's theories to the violent storming of the Capitol, we can view this as an attempt to storm an institution of materialised power by means of violence, in order to prevent the political act of installing elected decision makers in which the will of the voters, systemically derived through elections, is confirmed and institutionalised. This should be seen as a violent act against democracy.

As Freter (2019) further demonstrates, populism also uses verbal violence through language, and uses simplifications, emotions, and provocations to opportunistically win over the population by degrading politics. Since National Socialism contained many features of populism, populists try to distance themselves from it by emphasising the democratic features of their form of populism, such as people power. However, they only want to attract homogeneous people (ibid., 170f). According to Arendt's Theory, the anti-pluralism in populism vehemently contradicts democratic politics, where people are equals. Populism is, thus, not democratic.[42] Populists spread their views by stirring up fear and reading into, and sharing, false reports and conspiracy narratives. This indicates a proximity to demagogy, which relies on emotional manipulation and the stoking of prejudices (ibid., 171f).

Trust and Science

If scientific knowledge was based on social interest, or cultural values, then populists and conspiracy theorists could be correct in their views that there is no climate change, and the earth is flat. But science is not an ideology and it is not just influenced by social interest. Berger and Luckmann encouraged the concept of a dialectical process which would describe the relationship between knowledge and its social base. They claim that a dialectic process comes between the subjective stock of knowledge and the 'real' social world. Scientists are challenged by the 'real world'. For example, in the form of generalised others or by material resistance (where they gain knowledge based on scientific methods). Within the social world of science, there is a pre-existing corpus of scientific knowledge, established methods, and prearranged laboratories. The scientific environment and the world of conspiracy theories is quite different, but both arenas allow a form of performativity when people trust in them.

Research from the field of STS has shown how important communication of uncertainty of scientific knowledge is for establishing trust in the public sphere (Wynne 1992). Disagreeing with the PUS, Wynne (ibid., 282) argued that it is not an unwillingness or inability of the public to understand the correct scientific information, but rather a question of the public's trust in science and institutions socially. Based on a case study about sheep farmers in Cumbria—and their struggle with scientific experts and government advice after Chernobyl—he argued that:

> Understanding or knowledge, its precision and resilience, is a function of social solidarity, mediated by the relational elements of trust, dependency and social identity; constructing that intellectual understanding should be seen as process of social identity-construction.
>
> (ibid., 283)

Science is accepted in public based on the trust that people develop. In the case study, he reconstructed, based on interviews, how it happened that the sheep farmers lost their trust in the scientists and government, and partially developed kind of a conspiracy theory. Sheep farmers asserted that scientists announced conflicting statements about the contamination of their sheep several times in an arrogant style of communication, without ever mentioning previous failures. Additionally, the farmers disapproved of the fact that scientists ignored their local knowledge, and doubted the acumen of the scientists working on the field. By observing how scientists worked at their sites, they entered the black box of knowledge production. The expertise of scientists had a huge impact on the livelihood of the sheep farmers, who experienced an existential threat. A third source for the mistrust of the sheep farmers in science was the background of an unaccounted accident of a nuclear fuel reprocessing complex in Sellafield, next to their area. Lost trust in institutions from the past can lead to mistrust in institutions in the future. Even if Wynne's case study partly reproduces the Ethno-theory of sheep farmers about science, it shows us how conspiracy theories can develop and manifest in public. An acceptance of scientific advice in public can easily switch to hostility. Alternative constructions of reality are not only based on intellectual mistakes of recipients, but also on social failures of experts who are communicating (ibid., 293). In this case study, the farmers would have had a greater 'open-endess about scientific logical structures and its institutional and cultural forms than is usually recognized' (ibid., 295).

In the history of science communication, it was 'the generalization of trust beyond circles of people known to know one another' (Weingart and Guenther 2016) that created trust within institutions. Many researchers believe that modern science is anonymous, and that trust (Luhmann 1989) in faceless institutions (Giddens 1990) is key.[43] However, I disagree, and instead, rely on Shapin's work to argue against this one-sided perspective.

> So, one story about the modern condition points to anonymity and system-trust in abstract capacities, while the other identifies persisting patterns of traditional familiarity and trust persons. . . . One can characterize the modern condition through the serial application of both stories.
>
> (Shapin 1994, 415)

The study of the Science Slam does not just require trust in the system, but also trust in people. If we apply Goffman's (1981) thoughts about lectures to this, one could argue that institutional authority is justified in talks (and trust in the system is generated). If we take this view, science as an institution can be seen as being guided by trustworthy agents, scientists. This is only the case if scientists have a trustworthy image, and in the case of Science Slams, through their presentations. From the public's perspective, this works in two ways. If the author of the knowledge is trustworthy, this justifies the authority of the institution they are associated with, while in turn, the institution legitimises the author of the knowledge.

This is exemplified by my empirical study on Science Slams. What is emphasised in Science Slams is that the knowledge presented has to be genuine, scientific knowledge, discovered by the scientist themselves. Speakers at Science Slams are introduced as researchers, so audiences expect scientifically legitimate claims. In my extensive interviews with Science Slam organisers, I attempted to find out how they ensured that the content presented would be legitimate, scientific knowledge. Most of them said they simply trusted that the person giving the presentation would talk legitimately about science.[44] As most of the speakers at Science Slams are affiliated with a university, most people trust in their scientific legitimacy. This seems to be synonymous with earlier concepts of legitimacy. After the 18th century, for example, truth was no longer dependent on the person (mainly the aristocracy), but on the institution that the person was associated with. In the present day, Science Slam organisers sometimes confirm the legitimacy of a scientist by searching on Google for them to see if they are linked to a university. Sometimes, if applicable, organisers will read the scientists' publication list. If the list looks scientific (as opposed to pseudo-scientific), they assume that the scientist will speak legitimately about science. A further indication of validity is success at previous Science Slam events. Most of the organisers know each other and trust that someone who has been successful elsewhere will be a suitable candidate.

As for those in the audience, if they are not scientists themselves or part of the same discipline as the presenter, they will probably just have a feeling as to whether they believe that the speaker is a trustworthy scientist. The audience must trust both their general knowledge and their own personal feelings to decide if someone is legitimate. In my ethnographic research, I met some people who had been part of the audience at some Science Slams. I learnt that they had had discussions about the talks afterwards in order to evaluate them. Often, audience members assessed the talks based on their own personal feelings or knowledge. Some felt that the talks were about general topics rather than about the speaker's own research ('not much own research', 'it was entertaining but I did not learn anything new'). Informal Science Slam speakers are, therefore, judged on the basis of their performance on stage.

Shapin and Gieryn both questioned how scientific researchers can rhetorically and socially convince others to believe in certain knowledge. Modern sociological studies about the communication of scientific knowledge do not focus on interaction and embodiment. When analysing scientific communication, I look at the validity of scientific knowledge in the communicative genre of the Science Slam. The construction of something scientific occurs through the combination of immersed subjects, their forms of expressions (objectivations) and others. By using this, I will try to explain how the objectivity of reality is produced.

Notes

1 Daum's book *Wissenschaftspopularisierung im 19. Jahrhundert* (1998) about the popularisation of science in the 19th century tells how realists established the teaching of Biology in Prussian Schools in 1908 by means of a humanistic rhetoric. Meanwhile, natural scientists distanced themselves from engineers by using an anti-utilitarian rhetoric (for further details see Gieryn (1983)).Timothy Lenoir builds on this and argues that the German scientific entrepreneurs were so successful because they switched between utilitarian and humanistic rhetorics, depending on whether they were talking to new capitalistic bourgeois or humanistic bourgeois. For further details see Lenoir (1997).
2 The secret associations or 'associations of Enlightenment' in the 18th century can be viewed as instrumental in this movement because they were the first places where social and academic requirements of science were negotiable. Van Dülmen (1977, 252) describes how associations for the bourgeois were an important counterpart to courtly societies because they provided a space for bourgeois culture and thinking. Social interactions in these associations also had a huge impact on creating and preserving male bourgeois cultural values. Training consisted not only of 'sociability, education, critical thinking and self-affirmation' (Daum 1998) but also of scientific debate and self-improvement in natural science. Daum disagrees with the traditional approach of labelling the movement of popularising science primarily as liberal bourgeois and believes instead it came about through the work of professors. He points to the important role middlemen (like academics without certificates) also played in the movement.

According to Daum, many middlemen were publishers of magazines on popular science and founders of clubs on science so they had an important role in popularising science.

3 Alexander von Humboldt was one of the pioneers of national education. His famous cosmos-lectures about physical geography in Berlin (Sing Academy) in the year 1827 were some of the first scientific talks to attract a mass public audience. Humboldt aimed to talk to different social classes in order to spread knowledge about natural science. Thus, it was not only a bourgeois audience that listened to him, but also a non-academic audience (cf. Peters 2011, 62).

4 The assimilation of a person into a cultural environment through education.

5 He was not in favour of the tendency in public communication to rectify a lack of intimacy by creating an unreal secondary intimacy in the mass media. Habermas complained about yellow journalism and the 'human interest story'. For him, the growing number of citizens' private stories in the public sphere was associated with the end of high civilisation.

6 'Power corresponds to the human ability not just to act but to act in concert. Power is never the property of an individual; it belongs to a group and remains in existence only so long as the group keeps together. When we say of somebody that he is "in power" we actually refer to his being empowered by a certain number of people to act in their name. The moment a group, from which the power originated to begin with disappears, "his power" also vanishes' (Arendt 1970b, 143).

7 At the same time, the astonishment shown by audiences at scientific wonders was associated with a lack of knowledge because it was only people who were unfamiliar with cause and effect who were viewed as being able to be surprised (for example, according to Hume, astonishment was a characteristic of farmers, barbarians, women, and children).

8 Around the year 1820, the 'scientific persona' came into being. Science as a vocation had existed since the 18th century but at this time scientific concepts were always connected to practices of the mind and body (the self) like scientific observation, self-monitoring, keeping lab notebooks, quieting the will (Daston and Galison 2007, 198).

9 Examples are Dr Faustus, Prometheus, Zauberlehrling, and Frankenstein. Historically, the scientist has always been either naïve in his wish to save the world, or unscrupulous and obsessed in his ignorance and unaware of all the victims of his inventions.

10 'The Royal Society declared that it had realised Bacon's dream of joining a new science to a new social role for the man of science: not a professional scholar, not a Schoolman, not a slave to a philosophical system, not a professional cleric, and not a professional physician, but free, independent, modest, and virtuous seeker of truth about God's nature. Science, the Society said, had been remade into both a polite and useful practice, fit for gentlemanly participation and equipped to secure and extend state power' (Shapin 2006, 190).

11 Gender-oriented science studies have correspondingly described scientific performances as dissimilar to artistic, female, and amateurish representations (Heintz, Merz, and Christina 2004, 30). Sandra Beaufays and Beate Krais also show in *Doing Science-Doing Gender* (2005) that the academic world in Germany in the present day is still male-dominated because those who are viewed as manly succeed in science.

12 Jarausch (1990) explains that despite the fact there was a broad societal fascination for technical training, and that many engineers had a bourgeois background, middle-class intellectuals did not accept them as equal because they had relatively little classical humanistic education. This is not just true of Germany in the 18th century. Shapin tells us how even in the early modern era technicians were just the helpers of natural scientists and even in recent times, they had a lower working-class status and were often made to work backstage during experiments. For further information see Shapin (1989).

13 'At this time, male philosophers attempted to bring prestige to their occupation and institutions and to assert control and order over the natural world. In order to accomplish their aims, they believed they needed to ostracise women from the practice of science. As the 18th century progressed, women were increasingly being depicted as subservient and dependent members of society. As they lost their role as producers and were increasingly circumscribed to fulfil domestic roles in the home and as their status as reproducers was diminished in importance, women maintained their association with the natural world. As a result, men no longer believed women capable of making any substantial intellectual contribution to either science or society. The image of the goddess of Science was disappearing; no longer was Science a woman leading her followers along the path to knowledge. Moreover, Nature was increasingly represented as a willing woman, susceptible to the lure of men who wished to know her, unveiling herself before science. This metaphorical change highlights women's increasing absence from science as researchers, the decline of old traditions in science, and the rise of the new mechanical and experimental philosophy. But while women and their work, scientific or otherwise, were relegated to the shadows, the metaphors associated with women haunted institutions such as the Royal Society and scientists such as Sir Isaac Newton' (Sheffield 2004).

14 As Sheffield (2004) writes, Laura Bassi (1711–1788) received a doctorate in philosophy from the University of Bologna in 1733 and became a celebrated professor of physics and a member of the Academy of Sciences. However, at a similar time, astronomer Maria Winkelmann's request to join the Berlin Academy was turned down.

15 A good example of an ideological scientist is Konrad Lorenz.

16 'Marie Curie practised physics and chemistry in the first half of the 20th century, when these disciplines were still very much male enclaves. This introduction will follow her life achievements in science to outline the struggles, constraints, and barriers that she faced and to explain how and why she became the iconic woman scientist' (Sheffield 2004). Furthermore, female scientists were often posthumously portrayed as cold and distant.

17 Researchers wonder why it became popular again after the late Weimar Republic and the 1970s.

18 More about that topic can be found in Nowotny et al. (2001).

19 'It is, of course, no mere coincidence that the period of the rise of the entrepreneur type also gave birth to Utilitarianism' (Schumpeter 2000, 69).

20 Accessed January 15, 2014, http://royalsociety.org/uploadedFiles/Royal_Society_Content/policy/publications/1985/10700.pdf.

21 More about the history of PUS can be found in Gregory and Miller (2000).

22 Bucchi (2008, 58) outlined five characteristics with which to identify a concept of science communication. The first characteristic focused on the idea that the media acts as a channel designed to convey scientific notions. The second characteristic is the idea that the public is imagined as passive. The third is that science communication is a linear, one-way process. The fourth is the notion that communication is concerned with the transfer of knowledge from one subject, or group of subjects, to another. The fifth characteristic is the idea that knowledge is transferrable without significant alterations from one context to another.

23 The Science Slam was founded in 2006, in the German town of Darmstadt. The psychologist Alexander Dreppec, who came up with the idea for the event, was heavily inspired by the Poetry Slam event.

24 As well as multiple personas that describe scientific identities, I also found evidence of an "Engineering man" persona (Scientist as developer) in my research.

25 'A scientific persona is a collective identity that does not necessarily correspond with that of an individual, but shapes the aspirations, characteristics, lifestyles and even physical abilities of a group that is committed to this identity' (Daston 2003).

26 I aim to outline some characteristics and elements of popular scientific personas in contemporary public science communication. In this way, I would like to show certain elements that are common to most scientific personas, as they appear in Science Slams. Most Science slammers exhibit some of the described elements, but none of them completely fulfill all the characteristics of one particular type of persona.

27 'It means something else, namely, the knowledge or belief that if one but wished one could learn it at any time. Hence, it means that principally there are no mysterious incalculable forces that come into play, but rather that one can, in principle, master all things by calculation. This means that the world is disenchanted. One need no longer have recourse to magical means in order to master or implore the spirits, as did the savage, for whom such mysterious powers existed. Technical means and calculations perform the service. This above all is what intellectualization means' (Weber 1946, 129).

28 Other examples show that emotions like melancholy were included in the catalog of deadly sins in the Middle Ages under the name 'acedia' (Wiebicke 2010).

29 He stated that individualisation has increasingly led to a mass detachment of individuals from traditional ties. Since individual choice is a standard part of modern freedom, the individual has to choose geographical community membership options, beyond quasi-natural socio-moral milieus.

30 In wealthy societies, it is not survival that is the main cause of action, but rather the drive for experience.

31 What Kaerlein (2018) believes to be a problem is that smartphones have become a highly personalised medium, but they are still part of a media infrastructure that cannot be grasped (the smartphone is constantly connected to an unknown technical infrastructure). The high level of intimacy and the strangeness of technical environments stand in conflict with each other.

32 Mikko Salmela and Christian von Scheve (2018) have analysed how 'right-wing populism is characterized by repressed shame that transforms fear and insecurity into anger, resentment, and hatred against perceived "enemies" of the precarious self'.

33 To acknowledge these changes, I have adapted the following quotes: What is 'real' to a scientist may not be 'real' to an **American president** . . . . It follows that specific agglomerations of 'reality' and 'knowledge' pertain to specific social context, and that their relationships will have to be included in an adequate sociological analysis of these contexts (Berger and Luckmann 1967, 3, changes M.H.).

   American Presidents may be equipped with knowledge that is wrong from the standpoint of science (e.g., the view that there is no global warming), but this 'wrong' knowledge can guide actions—with very real results (rise of the sea game angle) (Knoblauch 2014, 158, changes M.H).

34 This might be a contrast to the post-democracy movement and against the apathy of citizens.

35 Accessed January 2, 2021, www.youtube.com/watch?v=rhQVustYV24 Greta Thunberg on the Daily Show.

36 Peters, Hans P., "Science Dilemma: Between Public Trust and Social Relevance. Public Mistrust Stems from Science's Ties to Economic and Political Interest," *EuroScientist*, 2015, accessed January 13, 2021, www.euroscientist.com/trust-in-science-as-compared-to-trust-in- economics-and-politics.

37 In this paper, I will take the perspective that methodological agnosticism (epistemological agnosticism) with its interest in what is true in various contexts, is right. I will try not to make any judgements about the truth of knowledge and will aim to show what practitioners do in their own right. The sociology of knowledge looks for social consequences of knowledge. As opposed to everyday life or philosophy, the sociological notion of knowledge is not directly related to truth. For some it may bring relief to bear in mind that from a socially constructive point of view, the constructed reality of

science deniers will come back to bite them. The truth will rise and Trump's golf court in Florida will disappear. There will always be a robust reality that cannot be wished away.

38 'Just as industrial civilization flourished at the expense of nature and now threatens to cost us the Earth, an information civilization shaped by surveillance capitalism and its new instrumentation power will thrive at the expense of human nature and will threaten to cost us our humanity' (Zuboff and Schwandt 2019, 25).

39 'Benkler et al. (2020) show that the narrative of supposed election fraud did not need digital media to overgrow the political landscape. It has been part of the political campaign of an elected president of the USA, who has ample access to traditional media. Accordingly, the narrative found widespread resonance in traditional media. Any analysis which only focuses on digital media and ignores the non-digital factors will fall short in actually understanding the backstory of the incidents' (Benkler et al. cited in Rau 2021). Source: accessed January 2, 2021, https://janrau.com/blog/.

40 In her book on totalitarianism Arendt describes the birth of a completely new form of state in which unprecedented crimes against humanity are committed. For Arendt, this is made possible by the combination of a leader in principle, a modern, bureaucratic, mass society and terror. In this state, responsibility and guilt are de-personalised. The state is like a 'domination of nobody' because no one person sees himself/ herself responsible for any actions. According to Arendt, the administrative apparatus of the Holocaust was run not only by sadistic fanatics, but also by affectless bureaucrats. These bureaucrats related their values and morals to the prevailing norms as if they were trivial things like table manners. Totalitarianism and extreme evil build on extreme fanaticism, and also on extreme relativism. When bureaucracy and totalitarianism meet, political principles are replaced by functional processes and an administrative apparatus. In the worst case, it becomes 'tyranny without tyrants'. Arendt's characterisation of the bureaucratic apparatus sees it as a machine, not a politically accountable administration of public affairs. Totalitarian regimes aim to make people totally controllable. For Arendt, obedience in a machine is prohibited.

41 Violence is hostile. Violence can enforce obedience and is very efficient in achieving short-term goals. Violent people reject plurality. In contrast to power, violence does not require a group of people but can be exercised by individuals and requires only a justified purpose in addition to means. Violence seeks to destroy power (Arendt 1970a), so if a state uses violence to preserve itself, it only does so at the price of its own power. If a revolution removes a state by force, the state cannot gain power by using violence (Arendt 1965).

42 Arendt argues that thinking is necessary to maintain politics.

43 This also allows people to trust in political institutions.

44 'Of course, we cannot control if the presented information is true, we cannot verify it. In general, our main target group is PhD students who present their research. There might be scientists with different opinions but this has a scientific value. And any publication can be controversially discussed after publication and this is good for science. I do not claim that some sort of ultima ratio should be presented on stage, or the last word. I cannot have this aim. Firstly, even if I had a team of experts from each department, this would not work. Therefore, I accept the fact that if a master's thesis has survived and their PhD has been recognized, or a post-doctoral has had work published, I suppose the content is good enough for us. And when the presentation causes a controversial discussion this can be transferred from to the slam to the pub. If the audience is willing to discuss with the Slammer I think that's super' (MS#44).

3 Developing a Theoretical Framework in Which to Study Science Communication

This chapter will focus on social constructivism. More specifically, on what social constructivism was like in the 1960s, when it began. I will also look at how the work of Goffman (1981) and the new approach of communicative constructivism (Knoblauch 2013) can improve the 'social' (Bloor 1976; Barnes 1974; Collins 1983) and 'practice' elements (Latour and Woolgar [1979] 1986; Knorr-Cetina 1988; Lynch 1993) in science studies. Due to the fact that knowledge is both the point of departure and the product of communicative action, we will also focus on the dual elements of knowledge and communication in this chapter. By using a three-sided perspective towards science and science communication, the communicative genre of the Science Slam will be outlined.

Science

Early studies described science as a field driven primarily by intellectualism. Since the 1970s, the fields of science and technology have both experienced a development in a number of social explanations which highlight how the understanding of science has changed. This chapter aims to convey one such perspective towards (scientific) knowledge, based on the assumption that there is a dialectic process between subjective knowledge and an objectivated social world. This section seeks to explain the relationship between STS and sociology (with respect to the sociology of knowledge), whilst also laying the groundwork for the proper method of the study of knowledge and science communication to emerge.

Science studies were in their early days, closely connected to the sociological debate (Merton 1973), yet they swiftly moved away from this when the philosophers took over (Bloor 1976). The debate within STS on this very topic, however, only partially recognises sociological theory. While many of the classical figures (Marx, Mead, Weber) at least included the possibility for dialectical thinking in their work, SoK researchers largely search for causal reasons to explain knowledge. As such, it seems as if STS's treatment of classical sociology sometimes re-enacts old debates. That is to say that STS researchers seem to have picked up questions from classical sociology while, at the same time, applying an interdisciplinary perspective. This perspective has given sociologists great case study examples.

DOI: 10.4324/9781003172635-3

In order to gain a greater understanding of the German perspective on the sociology of knowledge, I will now demonstrate some key features of social constructivism. In STS, there are ambiguous opinions as to what the connection between knowledge and its social base is and entails. Despite this ambiguity, a large majority of STS researchers argue that a correlative understanding of knowledge is best. For this reason, I will outline a genealogy of social constructivism. I will aim to express how a triple combination of the conceived (ideality), the perceived (physicality), and the performance of science can be viewed from the perspective of social constructivism and I will argue that this three-sided perspective on knowledge is the best approach. In this perspective, (scientific) knowledge is neither the result of social interest, nor of cultural values. Instead, I believe that the imagined 'ideal' and the physically or mentally 'perceived' have influenced the development of (scientific) knowledge. I will show how this triadic perspective, when applied to methodology, can enrich the way we look at science communication and how it can be useful in the study of new communicative genres like the Science Slam. As my small overview of classical philosophy and sociology shows, knowledge has always been a controversial topic. The social constructivist perspective in STS is no exception, for it too has created problems for public science communication.

Sociology of Knowledge

In the 1960s, Berger and Luckmann claimed that the sociology of knowledge had to deal with an empirical aspect, particularly in the study of knowledge in human societies: 'processes by which any body of "knowledge" comes to be socially established as "reality"' (Berger and Luckmann 1967, 3). Influenced by Marx's ideas on the relationship between material power and conceptual success, Berger and Luckmann suggested that he or she 'who has the bigger stick has the better chance of imposing his definition of reality' (ibid., 109).

> Two societies confronting each other with conflicting universes will both develop conceptual machineries designed to maintain their respective universe. . . . The historical outcome of each clash of gods was determined by those who wielded the better weapons rather that those who had the better arguments.
>
> (ibid.)

In the fight between the classical 19th century controlled, distanced, and emotionally reserved male bourgeois scientist and the popularised, approachable, and funny scientist of today, it is hard to know which one has the 'bigger stick'. It is important to know this because, according to Berger and Luckmann, the 'bigger stick' is necessary if one wishes to impose one's definition of reality. On the one hand, the material that constructed the universe maintaining legitimisations was seen as part of the expansion of power relations (ibid.). On the other hand, the machineries of universe maintenance were themselves described as products of social activity that created some sort of performativity. If we believe this perspective, the scientific universe not only has to deal with the difficulty of 'keeping out the outsiders' (ibid., 87), but also with the challenge of gaining acceptance so certain scientific procedures can take place. This is due to underlying tensions within public science communication, based on scientists' supposed privilege and on the general public's view of scientists. To counteract this, there are certain intimidation techniques that scientists use in order to draw boundaries between themselves and general society. Berger and Luckmann describe how scientists in the past asserted their authority by using age-old symbols of power and esoteric signs, such as outlandish costumes and incomprehensible language. Scientists also used professional knowledge and legitimising techniques to distinguish between scientific proof and quackery. Pierre Bourdieu in *Homo Academicus* (1988) explained the different ways one could use existing principles of social stratification. According to Bourdieu, some people have fewer ways of objectifying their symbolic capital and enforcing their perspective than others (cf. ibid., 49). Although Bourdieu took a somewhat antagonistic perspective, he argued that some people used symbolic power to change their social position. He likened the scientific scholar to a judge who had to decide whether people were scientific experts or not. This was done by using scientific instruments of objectivation, 'scientific weapons' and 'scientific effects' (ibid., 51).[1]

Science and Technology Studies

STS researchers have long pondered the question of how the powers of science can relate to scientific visual tricks. In *Laboratory Life: The Social Construction of Scientific Facts* ([1979] 1986), Latour and Woolgar described a 'circle of credibility' which scientists used to gain acceptance. Latour subsequently demonstrated that material differences help modern scientists convince others of their credibility. In the *Pasteurization of France* (Latour 1993), Latour gives an overview of how useful the anthrax vaccine was for communities in the French countryside. He further argued that Western science is so powerful today because it uses a gigantic scientific instrument, a panopticon, which allows scientists to produce optical consistency. So, the great divide or difference between Western scientists and others is that they use inscriptions as immutable mobiles in order to convince others.[2] I have already outlined how science communication and the Western scientific persona (Daston and Galison 2007) has changed over the years. These arguments will be the groundwork for my analysis of communicative construction in public science communication.

Despite STS researchers' suggestion that STS should be the leading source for public insight into science in action (Yearley 1994), a large amount of research is still needed in the fields of interaction, materiality, aesthetics and emotion in science communication. The issue in STS, of science communication being disembodied (Davies 2009) has been recently addressed by scholars.

Exceptions to this issue show that representations and visibility in science are generated and interpreted by social actors in the context of communicative processes (e.g., Lynch 1988; Amann and Knorr-Cetina 1990a, 1990b; Beaulieu 2002; Alac 2008). Additionally, a handful of scholars have focused on kinaesthetic and affective entanglement in science communication (Myers 2012, 2008), and on aspects of body movement and socio-technical arrangements in interaction (Goodwin 1981, 1994; Goffman 1981; Knoblauch 2007; Kiesow 2014; Tuma 2012). Yet, despite this, the various processes by which a body of knowledge comes to be socially established as 'reality' still needs more research, since such processes are embodied processes. My project aims to address these embodied communicative processes and to also focus on the vastly understudied area of public science communication. Although Science Slam organisers maintain the critical discourse on nonconformist science, it is difficult to break with dominant traditional Western ideas about science communication, as we will see.

The interdisciplinary field of STS seems, partially at least, unclear about its relationship with sociology and the sociology of knowledge. This ambiguity is due to several reasons. Firstly, only a small number of the STS founders were professionally trained sociologists. On top of this, there is a disagreement between some within STS as to the task, methods, and restrictions of the field (cf. Shapin 1995, 296). Thirdly, many STS researchers embrace different sociological traditions and hold different aims. Despite these issues, STS researchers do generally hold the same opinion about a few key areas of social constructivism.

> For STS, social constructivism provides three important assumptions, or perhaps reminders. First is the reminder that science and technology are importantly social. Second is the reminder that they are active—the construction metaphor suggests activity. And third is the reminder that science and technology do not provide a direct route from nature to ideas about nature, that the products of science and technology are not themselves natural.
>
> (Sismondo 2010, 70)

In the 1970s, the authors Bloor and Barnes (Bloor 1976; Barnes 1974) established a social constructivist perspective on science that referred to Thomas Kuhn's work. However, their very flexible use of the term 'construction' resulted in Kuhn, who was a central figure in the Sociology of Scientific Knowledge (which will be referred to from now on as SSK), distancing himself from SSK. Indeed, Kuhn claimed that SSK was an example of 'deconstruction gone mad' (Kuhn in Shapin 1995, 294). To understand what happened to social constructivism within the context of STS studies, I will give an overview of the theoretical lines and national trends of social constructivism in France, America, the United Kingdom, and Germany. The primary focus is to develop a heuristic-based study of knowledge and to explain the relationship between STS and sociology with respect to the sociology of knowledge.

Concerns About Social Constructivism

The term 'social constructivism' in the context of STS is thought to have its origins in the work of Berger and Luckmann (1967).[3] However, as we will see, this is only partly true. Berger and Luckmann's original characterisation of knowledge stated that knowledge is the 'certainty that phenomena are real and that they possess specific characteristics' (ibid., 1). In this sense, knowledge was thought of, and studied as, a 'real' phenomenon that constantly picked up objectivated social reality and reproduced it.

> It will be enough, for our purposes, to define 'reality' as a quality appertaining to phenomena that we recognise as having a being independent of our own volition (we cannot 'wish them away').
>
> (ibid.)

The analysis and understanding of social constructivism since 1967 has been very diverse, not only within STS, but also in sociology. With Berger and Luckmann's introduction of social construction in sociology, many forms of constructivism and constructionism were established. Despite the almost universal agreement that Berger and Luckmann were the first ones to develop social constructivism, there are historically many other descriptions of the term. The American-oriented description of social constructivism comes from Andrew Abbott (2001), who recommended distinguishing between three forms of constructivism: consensus constructivism (Thomas, Park), ideological constructivism (Marx, Mannheim, Goldmann, and Hauser), and structuralist constructivism (Levi-Straus, Saussure).[4] Abbott argued that ideological constructivism was more prevalent in European history than consensus constructivism (cf. 63). It is with Abbott's description of the history of social constructivism as a whole that problems arise, however. Although he noted that Berger and Luckmann initially introduced the term social constructivism to American readers, and although he recognised that they combined the ideas of Schütz, Mead, Marx, Mannheim, and Freud, he maintained that his distinction between the three forms of constructivism was better.

For the purpose of this chapter, I would like to offer a slightly different view on social constructivism by suggesting that, in Anglo-Saxon countries, discourse has been dominated by the constructionism established in the work of Thomas, Park (1928), Harré, and Gergen. This idea was previously addressed by Jo Reichertz who argued that in Anglo-Saxon countries, constructionists aimed to distance themselves and their work from social constructivism, because they believed Berger and Luckmann had incorrectly focused solely on the individual mind and on single subjects (cf. Reichertz 2013, 62). If this perspective is correct, Shapin's speculation that the shift to social constructivism within STS was not driven by phenomenological approaches but by theoretical sources (e.g., labelling theory), is probably right (cf. Shapin 1995, 296). In Europe, an additional form of constructivism called radical constructivism (Glasersfeld, Maturana & Varela, Luhmann) was also introduced. This form was inspired by systems theory and neurobiology. Radical constructivism in Europe was different from Berger and Luckmann's social constructivism because it was not very interested in the reified world. For this reason, in Germany, Berger and Luckmann attempted to save social constructivism from misinterpretation as being a theory of randomness (the mantra being: 'let's construct a reality') (cf. Loenhoff 2011). Luckmann also clarified that the term construction was only selected in order to distinguish it from the 'constitution' term in phenomenology. To highlight the dissimilarity between different forms of constructivism, Luckmann once stated that he 'is no constructivist' (in a radical sense), but that he believed in a realistic ontology and epistemology (cf. Luckmann 1999, 17).

This is reminiscent of internal debates in STS and of its current 'turn to ontology' (Lynch 2013). Ontological theories in STS aim to describe how realities are made or enacted in practices based on empirical observations. Not all agree that an ontological focus in STS is correct, however. Aspers (2014) stated that it was unclear what STS gained in arguing over ontology (or metaphysics). According to Aspers, empirical observations refer to constructed elements, so 'there is no fundamental qualitative difference between the ontological turn and what we know as constructivism' (Aspers 2014, 1). Michael Lynch added that contemporary STS risks straying into philosophical grounds when it focuses on ontology and on searching for prior assumptions about the world. Lynch argued that many have incorrectly researched ontology in a way comparable with idealist philosophy.[5]

Contemporary debates in STS (like the ontology debate) are historically grounded in sociology and the sociology of knowledge. When Berger and Luckmann argued against focusing on ontology in the 1960s, they based their arguments on Marxism. They were also interested in the commodity character of human activity.

> The process by which the externalized products of human activity attain the character of objectivity is objectivation. .... In other words, despite the objectivity that marks the social world in human experience, it does not thereby acquire an ontological status apart from the human activity that produces it. The paradox that man is capable of producing a world that he then experiences as something other than a human product will concern us later on.
>
> (cf. Berger and Luckmann 1967, 60)[6]

Social constructivism was originally characterised as a set of evidence that wanted to explain how the 'objectivity of reality' (Knorr-Cetina 1989, 89) was produced. Berger and Luckmann's social constructivism maintained that there is a robust reality that cannot be wished away. According to them, it is not the ontological perception of nature that is of interest, but rather the ontological status of social reality, which they explained by referring to history. Ultimately though, their task was not to define prior assumptions about the world, but to ask why members of society experience institutions or things as ontological.

As one aim of this chapter is to explain the relationship between STS and sociology with respect to social constructivism, I will now mention some differences and some similarities between STS and sociology. In the following passage, I will outline a few ways in which STS has received social constructivism.

The Problem With Social Construction and Science Communication

In the last few years, British researchers have mixed and amalgamated the ideas of scholars such as Wittgenstein, Durkheim, Marx, and Kuhn[7] in order to develop a new empirical approach to the study of science.[8] The re-evaluation of science as a social activity, as in research done at the universities of Bath and Edinburgh, has created problems for legitimacy in science. By focusing on the similarities between scientific knowledge and other forms of knowledge, the line between expert and non-expert has blurred and thus the need for science studies has also reduced. An example which shows how experts tend to ignore the knowledge of non-academics comes from Wynne's (1992) remarkable paper on sheep farmers in Cumbria. His work shows how the knowledge of sheep farmers was long ignored by scientific experts after a nuclear power plant accident. In the 1980s, Harry Collins argued that science should be seen as a procurer of certainty in relation to the extent it is distanced from the research front.

> That is to say, science only looks certain when one moves away from the 'core-set', either in sociometric space or time.
>
> (Collins 1987, 692)

If we agree with this assessment, disconnection to the production of scientific findings creates a different way of handling certainty. A typical feature of science communication is that non-existing certainty disperses through public presentation, without communicating the production of knowledge. This disconnection has been thought of as the reason why science is untouchable in communicative processes. Bruno Latour and Steve Woolgar ([1979] 1986) used terms like 'splitting' and 'literary inscription' to describe similar processes. In these processes, the production of 'facts' was deleted from the scientists'/recipients' consciousness (in Marx's words, this could be described as commodity character). In produced texts and in public presentations by scientists, the certainty of academic practices is overstated while controversies and failures are excluded. Shapin and Schaffer (1985) showed, with examples, that facts become mobile and immutable once they are displayed in the public sphere. The public often had, according to Yearley, idealised notions about scientific activity (cf. Yearley 1994, 245) and a vague idea of the scientist's work. Yet, on the subject of scientific expertise, even sceptics had contradictory arguments. In the 1990s, STS explored how their findings were used by non-scientific people. This was explored by Latour in his paper *Why Has Critique Run Out of Steam?* (2003). He argued that social constructivist critiques towards science may be inappropriate and have unintended consequences.

With this in mind, he also wondered whether, in society today, the danger may not lie in ideological arguments that are dressed up as facts, but in mass disbelief in supposed facts, which are then turned into ideology; for example, global warming and climate sceptics used arguments from social constructivism. To certify the work order of STS Latour stated:

> The question was never to get away from facts but closer to them, not fighting empiricism but, on the contrary, renewing empiricism. . . . the critical mind, if it is to renew itself and be relevant again, is to be found in the cultivation of a stubbornly realist attitude—to speak like William James—but a realism dealing with what I will call matters of concern, not matters of fact.
>
> (Latour 2003, 231)

Those who are involved in science studies today are aware of the consequences that the deconstruction and loss of internal and external communicative bases could have on the legitimacy of scientific projects (Collins and Pinch 1993). If a scientist's theories and observations influence what he sees in (and as) data; if there is no clear way to develop a theory from data; if scientific practice is always influenced by its historical context, then how can one even begin to start or develop research that is unbiased? This self-induced paradox of the sociology of knowledge similarly confused SKK, 'What is the objective base of sociology of knowledge if legitimacy of knowledge is not just an epistemic question, but also a question of social constellations that make judgements about validity?' (cf. Schützeichel 2012, 18).

To avoid getting lost in incommensurability and cognitive relativism, and to enable effective communication between disciplines, some feminist researchers have tried to define some universal epistemological standards (like empiricism) for heterogeneous scientific fields, through their search for a 'successor science' (Harding 1986). Harding, for example, identified the need for an 'earth-wide network of connections, including the ability partially to translate knowledge among very different—and power-differentiated—communities' (ibid). Additionally, Haraway argued that a feminist objectivity is 'about limited location and situated knowledge, not about transcendence and splitting of subject and object' (ibid., 583).[9] For her, the solution to the problem of objectivity was to look at the context the knowledge was situated in.

> So, I think my problem, and 'our' problem, is how to have simultaneously an account of radical historical contingency for all knowledge claims and knowing subjects, a critical practice for recognizing our own 'semiotic technologies' for making meanings, and a non-nonsense commitment to faithful accounts of a 'real' world.
>
> (Haraway 1988, 579)

Haraway described positioning as 'a key practice', as opposed to the typical visual tricks and powers of modern science which makes 'various forms of unlocatable, and so irresponsible, knowledge claims' (ibid., 583). In general, these feminist approaches suggest that the focus should be on embodiment, partiality, and localisation as grounding practices for knowledge claims (objectivity).

On the other hand, Harry Collins and Robert Evans (2002) argued that a 'Third Wave of Science Studies' (2002) was necessary to hold back the 'implosion' (Shapin 1995, 311) of SSK. They tried to reconstruct knowledge that had been deconstructed by the 'social turn' of science studies by the use of a normative theory of expertise.[10] Their aim was to bring different kinds of expertise back onto the agenda. Collins, for example, argued that there should be experts of certain fields of knowledge. Yet, what he called 'relativistic' research, under the umbrella term 'Wave Two', was blind to other approaches of science studies.[11] Lynch, meanwhile, was critical of the fact that researchers like Collins and Evans were turning away from descriptions of scientific practices, and turning instead to a more normative engagement with science. At the same time as this, generalist and essentialist theorising in STS was condemned. Although social processes (for example, finding a census in the laboratory, processes of 'literary inscription', reducing scientific claims, construction of the laboratory) were identified as essential for 'science in the making', the social itself did not become a social interest. This was because science studies in general focused on two things: a scientist's work and explaining what scientists do. Shapin suggested making SSK the leading organisation for public insight into science in action. Others argued that genres in public science communication could learn a lot from STS, because STS was the perfect field for introducing the public to science in the making (cf. Yearley 1994, 245).

> I have argued that science studiers have offered a coherent and empirically rich understanding of what the doing of science is like, what the character of scientific knowledge is, and what the typical dilemmas of the public use of the scientific authority are.
>
> (cf. Yearley 1994, 253)

Despite such views, the social constructivist perspective is a problem for public science communication. I will now demonstrate some key features of social constructivism within the context of STS. In STS there are ambiguous opinions as to the connection between knowledge and its social base. Many researchers argue for a correlative understanding of knowledge and for this reason, I will also outline the genealogy of social constructivism. After that, I will explain why I think communicative constructivism is a useful aid in the development of research of the social constructivist programme. I will then relate these thoughts to contemporary developments of post-truth politics and to the way that trust is built in different societies.

Social Constructivism

Peter L. Berger and Thomas Luckmann first met each other in the 1950s at the Graduate Faculty of Political and Social Science in the New School of Social Research in New York. The New School was established by Columbia University with the aim of offering academics, who were forced into exile by the Nazi regime, an opportunity to continue their research abroad. When Berger and Luckmann happened upon each other for a second time at the New School, as professors in the 1960s, they decided to work on a social theory which was inspired by the sociology of knowledge and based upon philosophy and anthropology. Their work, *The Social Construction of Reality*, was first published in 1966 and, among other things, it was a reaction to the then popular structural functionalism of Talcott Parsons. Unlike Parsons, Berger and Luckmann wanted to focus on the social relativity of knowledge and reality.

> What is 'real' to a Tibetan monk may not be 'real' to an American businessman. . . . It follows that specific agglomerations of 'reality' and 'knowledge' pertain to specific social context, and that their relationships will have to be included in an adequate sociological analysis of these contexts.
>
> (Berger and Luckmann 1967, 3)

In contrast to realists who argue that 'truth is more dependent upon the natural world than upon people who articulate them' (cf. Sismondo 2010, 58), social constructivism relied on the influence of the 'natural world' and also treated forms of knowledge as real social research objects. Berger and Luckmann's methodological agnosticism (epistemological agnosticism), aimed to discover what was seen and acknowledged as true in different contexts.

> It is our contention, then, that the sociology of knowledge must concern itself with whatever passes for 'knowledge' in a society, regardless of the ultimate validity or invalidity (by whatever criteria) of such 'knowledge'. And insofar as all human 'knowledge' is developed, transmitted and maintained in social situations, the sociology of knowledge must seek to understand processes by which this is done in such a way that a taken-for-granted 'reality' congeals for the man in the street. In other words, we contend that the sociology of knowledge is concerned with the analysis of the social construction of reality.
>
> (Berger and Luckmann 1967, 3)

Berger and Luckmann's approach is similar to the ethnomethodological position taken by STS researchers Lynch and Woolgar. A typical aspect of both ethnomethodology and social constructivism, for example, is that both try to reserve judgement about the truth of knowledge. Berger and Luckmann aimed to show what practitioners did in their own right, rather than showing the breakdown of social constructions. They did not view knowledge as inherently wrong or right but looked instead at the social consequences of the knowledge. For them, truth was validity, and validity was socially established (cf. Knoblauch 2014).

> As opposed to everyday life or philosophy, the sociological notion of knowledge is not directly related to truth. Actors may be equipped with knowledge

that is wrong from the standpoint of science (e.g., the view that the earth is flat), but this 'wrong' knowledge can guide actions—with very real results.

(ibid., 158)

Unlike in conflict theory, where judging differences is obligatory (and moral neutrality was certainly not supported), Berger and Luckmann attempted to generate a social theory where being impartial was key.

The Relevance of Socially Derived Subjective Knowledge

As Knoblauch highlighted, the correlative perspective on knowledge views knowledge more or less as an independent entity separate from society. In this approach, the two entities (knowledge and society) are separated in order for them to relate to each other. However, the integrative perspective on knowledge did not separate social structure and knowledge because knowledge was constitutive for social order and for the construction of reality (cf. Knoblauch 2005a, 18). Knowledge in this sense was not only related to previous action but also appeared mostly coupled. In this way, the work of Weber on social action, Mead on viewing knowledge as integrated into the social, and Schütz on the sedimentation of knowledge were all sources of inspiration. A correlative perspective on knowledge can still be viewed as important if we look at the institutionalised stock of knowledge in our society that is used in day-to-day experiences. Other work such as the 'taken-for-granted-facts' described by Fleck (1980), Mannheim's 'Denkstil', Kuhn's (1970) paradigm, and Knorr-Cetina's epistemic cultures have all explicated a historic and institutionalised perspective on knowledge. At the same time, the process of knowledge production is the focus of constructivism. So, Berger and Luckmann's approach had an understanding of knowledge that took into account the description of institutionalised knowledge.

The knowledge of members of society is divided into different areas. Most subjective knowledge is 'general knowledge' because the lifeworld of everyday life is the dominant reality. Most members of a society cannot answer the question of what a 'science slam' is, so this knowledge does not belong to general knowledge. General knowledge as socially shared knowledge is contrasted with the special knowledge of specialised, often institutionalised areas of knowledge. General knowledge includes routine knowledge (e.g., walking), practical knowledge (e.g., reading), and receptive knowledge (automated knowledge elements adopted from others).

The typifications acquired through experience are institutionalised to varying degrees. They are controlled by relevancies that are socially shaped, and have emerged through the activities of consciousness. In his essay 'The Stranger', Alfred Schütz makes clear how important the knowledge of cultural and civilisation patterns is in order to get along in a society. Members of society organise their knowledge of the social world according to its relevance for their actions (cf. Schütz 1972, 54). These relevancies are reflected in a 'system of knowledge' (ibid.: 57): 'It is a knowledge of trustworthy recipes for interpreting the social world and for dealing with things and people' (ibid.: 58). These recipes offer actors instructions for action on one hand, and schemes for interpretation on the other.

Following Schütz's study of the mundane, social constructivism described intersubjectivity as based on real life meetings with other people. If we believe Schutz, the 'lonely ego', described by Husserl in his theory of intersubjectivity, can be rejected.[12] Empirically, other people do exist before our birth, and communication with others is essential for the development of the self. This idea is similar to Mead's approach, in which the development of the mind and the self through communicative processes (role-taking) between organisms, and in relation to the objectivity of the world, is analysed. The subject, alter-ego relations and the object-subject relation all have their foundations in action. Knowledge, meanwhile, is built into interactions with people or things and has its origins in social processes (knowledge is socially learned). Following in the tradition of Weber, social action in this sense is described as the essence of the social. In other words, people orientate their actions to follow the actions of others—this is what sociality is (cf. Knoblauch 2013, 27). Sociologists of knowledge are fascinated by shared thoughts, actions, objects, and the relation of product, producer, and recipient within this. In studying new shared thoughts, actions, and objects, the difference between typification and existing knowledge becomes important.

The Relevance of the Objectivated World

For Hegel (1955), dialectic was a movement in consciousness that transformed everyday experiences into philosophical enlightenment (cf. Mikl-Horke 2001). This was a three-fold movement: from the subjective mind, over the objective mind, to the absolute mind. Dialectics describe this process as the mind recognising itself. Hegel believed that human beings were capable of building reality with a sharp contrast between thought and reality (*Gedanke und Wirklichkeit*). One could argue that this movement in consciousness was driven by contradictions. According to the theory, every positive thesis in the mind is confronted with a negation. This negation was then confronted with a negation. The second negation created a synthesis that made it possible to enter a higher ground of consciousness. With such a process based in consciousness, mankind was seen as being capable of stepping up to a higher form of mind (*Geist*). The key idea from Hegel's idealistic thinking was, however, the view that external reality is not self-sufficient, but guided by meanings and expressions of the mind which included and expressed the mental. People objectivate their thoughts and during this process, they see themselves as an objective mind. The mind absorbs the social, and the faculty of reason (*Vernunft*) becomes the organising principle of reality (cf. ibid., 38).

Marx amended Hegel's dialectic method. Marx's dialectic was similarly focused on a movement toward liberty, but instead of talking of a reality that could be found in the mind, he understood reality as material. Marx condemned the possible end of objectivity, which he believed to have happened in Hegel's philosophy.[13] Instead, Marx developed a material dialectic that aimed to highlight the contradictions of the objectivated world. Due to the fact that consciousness was seen as a consequence of economic relations, the dialectic movement was understood to be an economic process. Indeed, dialectics as a whole was a political and economic process that should have led to the removal of economic contradictions (cf. Mikl-Horke 2001). Instead of simply believing Hegel or Marx, Berger and Luckmann's social construction of reality considered the importance of both the material reality *and* the history of ideas.

> No history of ideas takes place in isolation from the blood and sweat of general history. But we must once again stress that this does not mean that these theories are nothing but reflections of 'underlying' institutional processes; the relationship between ideas and their sustaining social processes is always a dialectical one. . . . Consequently, social change must always be understood as standing in a dialectical relationship to the 'history of ideas.' Both 'idealistic' and 'materialistic' understandings of the relationship overlook this dialectic, and thus distort history.
>
> (Berger and Luckmann 1967, 128)

The ongoing institutional mode of selection that we see in structural functionalism (Parsons) or system theory (Luhmann) is simply not possible in the social construction of reality. For Berger and Luckmann, there is no history of ideas without blood, and no material social reality can develop without thinking people. This idea is important for, and relevant to, contemporary debates in Germany on this matter.[14]

Marx wondered why there was a commodity of human products and why people believed in the illusion that the value of the commodity came from within itself. In my opinion, STS, social constructivism, and communicative constructivism should all continue Marx's work by emphasising that there is a social construction of reality and that the rebuilding of the institutionalised parts is possible. This can be related back to Arendt (2017) and her reflections of totalitarianism and evil. In her lecture *Some Questions of Moral Philosophy,* Arendt (2006) pondered questions of ethics and morality and how we can reflect on people who act evilly. She challenged the views of moral philosophers like Hegel, Kant, Socrates, and Nietzsche and confronted them with the Holocaust. She was a participatory observer of the court proceedings against the mass murderer, Adolf Eichmann, a major figure in the Nazi regime. After attending Eichmann's trial she shifted her concept of evil from 'radical evil' to the 'banality of the evil' because she believed that ultimate evil was done by very simpleminded people who were driven by mindlessness (she called Eichmann 'Hanswurst'). Arendt understood Eichmann's inability to express himself in the courtroom as an inability to think and judge. In her opinion, his thoughtlessness allowed him to become one of the greatest criminals of all time as conscience arises from thinking.[15]

Although people often see themselves subject to inaccessible authority, similar to Franz Kafka's character Josef K. in his novel *The Trial*, they are nevertheless active members of this authority. Berger and Luckmann argued that the institutional world has no ontological stance, aside from the human activity that creates it. On the other hand, members of society experience institutions as inaccessible and ontological (ibid., 60). Arguably, institutions are human products but paradoxically seem non-human (for the materialistic version of this idea, refer to Marx's aforementioned ideas about commodity). Unlike in Marx's (Marx 1990; Marx and Engels 1968) arguments, thinking in society today is not solely derived from the most powerful class and is not therefore an ideal expression of powerful material relations. Although the biographical and social interests of people are reflected in their knowledge, it is still not an entirely determined category. Institutionalised spheres of knowledge have their own dynamics that back up onto members of society (e.g., in the case of socialisation). Social constructivism is not only interested in the dialectic process between the producer and his product, but also in how the product related back to the producer (cf. Berger and Luckmann 1967, 61).

> Society is a human product. Society is an objective reality. Man is a social product. It may also already be evident than an analysis of the social world that leaves out any one of these three moments will be distortive.
>
> (ibid.)

The dialectic process, according to Berger and Luckmann, is also relevant when individuals identify with other people (identification/self-identification); 'The individual not only takes on the roles and attitudes of others, but in the same process takes their world' (ibid., 132). Unlike in SSK, where the focus is on a one-sided societal influence, Berger and Luckmann described how a more distant relationship between a person's social interest and their knowledge could exist. Berger and Luckmann further described something that can be called a performativity of stocks of knowledge (their example of socialisation demonstrates this). According to this theory, two things are true. First, science is not solely a consequence of social interest, and second, if knowledge is viewed as social reality, it can create performativity when placed in an ongoing dialectic process with society: 'Definitions of the reality have self-fulfilling potency' (ibid., 128). Although Berger and Luckmann were mainly interested in everyday reality, they did also comment on the institutionalisation of scientific knowledge. When they highlighted that knowledge could be detached from the biographical and social interests of the knower, Berger and Luckmann argued that:

> In other words, the scientific universe of meaning is capable of attaining a good deal of autonomy as against its own social base. Theoretically, though in practice there will be great variations, this holds with any body of knowledge, even with cognitive perspectives on society.
>
> (Berger and Luckmann 1967, 88)

In describing the relationship between knowledge and its social base, Berger and Luckmann encouraged the concept of a dialectical process: '[K]nowledge is a social product and knowledge is a factor in social change' (ibid., 87). In other words, there is a dialectic process between the subjective stock of knowledge and the 'real' social world (this challenges us in the form of generalised others or a material resistance). Thus, scientific knowledge is not the causal result of social interest but is instead a 'real' social world. Within this social world, several factors or variables are not up for negotiation: the corpus of scientific knowledge, the laboratories worked in, and the colleagues worked with. Scientific reality is also only acted upon by scientists. Berger and Luckmann described three stages of dialectics:

A   Externalisation is the stage in which people build knowledge based on their (physical and mental) experiences of the world.
B   Objectivation is the stage where knowledge takes on an objective form (independent of people).
C   Internalisation is the stage in which people learn about the objective form and make it part of their consciousness.

More broadly, scientific beliefs or genres can be materialised by bodily and cognitive expressions (objectivation) and can then be internalised by others. Berger and Luckmann's dialectic process additionally highlighted that it was impossible to believe that materiality was created out of interest. Instead, the material reality was partly constructed through objectivation. This made it possible for knowledge to be independent from people. Objectivation, meanwhile, explained why people believed that society had a reality in itself (Marx's commodity character).[16] The problem was that people forgot that the social world was made by men and could be reconstructed. The non-human had a special type of legitimacy because it created facticity.

Studying Science as Communicative Construction

In a similar way to laboratory constructivism, the recent movement of 'communicative constructivism' (Keller, Knoblauch, and Reichertz 2013) has to a large degree tied the sociology of knowledge to materiality and empirical research. When it began in the 1990s, communicative construction was interested in the empirical research of communication. As a movement, it developed through Berger and Luckmann's ideas about the social construction of reality through the use of language and other forms of embodied social knowledge. Material signs, bodily attitudes, mimicry, gestures, objects, pictures, and diagrams became bodies of interest. Although communicative constructivism as a movement helped develop some of Berger and Luckmann's work, it also distanced itself from Luckmann's idea of 'proto-sociology' (the notion that philosophy is the untouchable truth upon which sociology is based). Due to this, communicative construction opened the door for a closer investigation of scientific knowledge. The empirical research of communicative constructivism additionally allowed for one to look at the communicative construction of scientific reality without the problematic 'turn to ontology'. As such, I now wish to outline the main cornerstones of this approach.

> The central idea of communicative constructivism is that everything that is relevant about social action has to be communicated/objectivated (observable and experienced).
>
> (ibid., 27 [translated by MH])

To begin with, knowledge, as the socially mediated meaning that guides action, is empirically observable (Knoblauch 2014, 156). The intended, and unintended, give-and-take effects of action ('*Wechselseitiges Wirkhandeln*' of Schütz) shape the communicative character of social action. Knoblauch emphasised the role that different types of objectivations that objectify meaning have.[17] In simpler terms, reality needs material manifestation. For example, talking leaves a different physical trace to dancing, punching, or writing.

> Objectivations may be, however, also objects, such as smoke signals, ink letters, or hieroglyphs. Objectivations also refer to objects to the degree that we relate to them in some active way.
>
> (ibid., 159)

If we look more deeply at knowledge and the different ways that objectivation can happen, for example, in conceptions (ideal meaning), actions (practices), or the perception of physical reality, we may miss the influence that different levels in the dialectic process can have. In general, within communicative constructivism, the key element (pivot) of communicative action is the body (the body has a voice, and a finger to push the button of a keyboard, for example). In this sense, communicative action can be described as performative. Communicative constructivism also has a triadic structure.

> Communicative action is based on a triadic structure: It is oriented to others, to an embodied subject, and related objectivations, that are seen as part of a shared environment.
>
> (ibid., 33 [transl. MH])

In the social process of communication, 'situated' actions (ethnomethodology) of transmitting knowledge are important, but so also are the durable 'situated' objectivations (transcending situations). Objects, like forms of technology, can be institutionalised objectivations because they create expectations, such as how to coordinate action (ibid, 38). So, the multimodality of communication is at the centre of empirical interest. This empirical interest in objectivation also means that communicative constructivism is interested in the temporal and spatial locations in which objectivations are sequentially realised (Body, Time, Space). The communicative constructivist perspective defines science as the communicative construction of something as scientific. This perspective, therefore, acknowledges the role that communication plays in the creation of knowledge.

A Plea for the Objectivated World

Since the 1970s, STS has argued that society should take into account the ambiguity and mess of science. Nearly 50 years of science studies later, it is clear that this also holds true for the interdisciplinary field of STS itself. In STS today, social constructivism is partly understood as a way of continuing on the idealistic tradition of Hegel. On the other hand, in Germany today, social constructivism is based on an integrative concept of knowledge, very different from German idealism. To demonstrate how different these approaches are to the radical constructivist line, Luckmann once said he was not a constructivist, but that he believed in an ontology and epistemology (Luckmann 1999, 17). Luckmann's social constructivism allowed for the possibility that there is a robust reality that members of society cannot wish away. The social constructivist perspective follows a theoretical line that works with an integrative concept of knowledge and also looks at the material 'objectivicated' aspects in the communicative construction of science and science communication.

As demonstrated earlier in this chapter, communicative action can best be viewed as a triple combination of subjective knowledge, the relation to others, and the objectivated world. This view is useful not just for scientific knowledge, but also for knowledge in different genres. The genres of science communication, meanwhile, are not purely the result of social interest, nor only relative to cultural values. Participants in science genres develop 'valid' shared knowledge and conceptions for science communication by relating to the perceived objectivated world. In Science Slams, institutionalised knowledge about a certain genre, and social relations between a presenter and an audience are all part of a reality that a scientist cannot wish away. In this section I have outlined how the new approach to communicative constructivism (KellerKnoblauch, and Reichertz 2013) allows the 'social' (Bloor 1976; Barnes 1974; Collins 1983) and the 'practice' (Latour and Woolgar [1979] 1986; Knorr-Cetina 1988, 1984, 1999) to develop in science studies. In this project as a whole, I ask why, and how, new communicative actions are realised in the midst of a complex and uncertain relationship between science and the public. In a later chapter on empiricism, it will become clear how the triadic perspective I have developed in this chapter can enrich the way we look at science communication, particularly in the study of new communicative events like the Science Slam.

The Shadows of Communication

In 1917, when Weber claimed that a scientist's personality should be characterised by their passion to serve science, he was lecturing to an audience of students about his work *Science as Vocation* (Weber 1977, 15). In this, he described lectures as one-sided apolitical talks, dissimilar to seminars or other talks, in which professors try to convince students to follow their judgements. He compared the young people at his lecture to the prisoners in Plato's cave allegory, for in his opinion, the students preferred to look at the shadow theatre on the cave wall instead of searching for enlightenment.

Much later, in the 1980s, Goffman lectured on a similar topic. According to Goffman, the printed text of a presentation is the most relevant part of a lecture and should therefore be the focus of attention. For him, the content of a lecture could conflict with the way the content was staged. Although Goffman believed that text was the most important aspect, he claimed that it often bored the audience and made their attention slip. The audience would focus instead on the way something was performed. In the struggle for attention between text and the staging of it, text often had a difficult starting position. For Goffman, therefore, presenting a scientific talk was a paradox, wherein the speaker often risked losing the content of the talk; 'Every transmission of signals through a channel is necessarily subject to "noise", namely, transmissions that aren't part of the intended signal and reduce its clarity' (Goffman 1981, 181). Like Weber, Goffman also described something similar to shadow theatre that prohibited listeners from truly understanding the content of a task. However, Goffman did not believe that we could, or should, get rid of these shadows.

In the Science Slam community, the equivalent of the shadow theatre, which can hold listeners back from truly understanding scientific content, is slapstick comedy. A blog-discussion within the Science Slam community questioned whether organisers should introduce stricter conditions in order to prevent Science Slams from becoming comedy slams.

Therefore, I plead for a kind of review or quality control of the presentations before a slammer is allowed onstage. One could start by clearly communicating terms and conditions, which should state that a Science Slam must deal with the research topic and should not be conceived as a comedy performance. One could also ask the potential slammer to fill out a self-assessment questionnaire, with questions like:

1   Please name and briefly describe your research topic.
2   What can the audience learn from your slam, and what message should they take home?
3   Please estimate (in percent) how much of your slam is based on facts and how much is based on gags and slapstick?[18]

Presentations in which the audience are 'mightily amused' but do not learn anything are believed to defeat the object and frame of the Science Slam. The distinction between facts and slapstick comedy in the Science Slam resembles the aforementioned shadow theatre.[19]

Communication Corrupted?

Fraser and Habermas's judgements about public discourse were related to the question of how communication could empower publics. As I have shown, the empowering nature of science communication takes a central role in STS. When relating this to new genres like the Science Slam, one could ask in what way communication contains a similarity that produces or propagates dissimilarity. Although we can ask this, I argue that, for methodological reasons, we should instead focus on how science is visible in public today and on how trust in science is generated. I further believe that the Science Slam genre should not be shunned as a phenomenon of cultural decay, without at least looking at it first.

Suggestions that new forms of science communication such as the Science Slam demonstrate cultural decay have permeated into many recent studies on the uptake of science in various fields of media. Like Habermas, Weingart (2005) in his study of mass media, argued that phenomena like the Science Slam marked a decline in the quality of communication about science. On the other hand, when comparing movies, scientific journalism, and literature, Haynes (1994) showed that the communication of science varied greatly according to which medium of representation was used. Criticisms aimed at the portrayal of science in the media mainly focus on the dominant role that the media in question has. Weingart talked about 'mediated science' because according to him the link between science and the public was dominated by mass media. Weingart also focused on the problems that could arise from the pressure of mediation. For him, the danger with relying too heavily on mass media was that the image of science could be slightly altered through such media.

Through the use of STS studies and from the sociology of knowledge, I have tried to highlight how complicated this perspective is. In the process of moving forward with social constructivism, the philosophy that I will focus, and base my arguments, on is methodological agnosticism (epistemological agnosticism). In short, methodological agnosticism is interested in what is acknowledged and seen as true in various contexts. As a social constructivist-based perspective, it further aims to withhold any judgements about any knowledge truths. Whilst trying not to show the breakdown of social constructions, methodological agnosticism strives instead to show what practitioners do and achieve in their own right.[20] It does not consider knowledge to be inherently wrong or right but looks instead at the social consequences of said knowledge. When analysing science communication, I will look at the validity of scientific knowledge particularly within the Science Slam genre. Old-fashioned ideas about communication are shown in several areas: science communication, politically based science communication programmes, and new forms of science communication. In the next section, I will show that it is necessary to have a concept of communication when discussing science communication.

Communication and the Self

In his work, Plessner labelled the distance between man and himself as 'condition humana'. The tension between a physical being (*Körper haben*, having a body) and the phenomenal body (*Leib sein*, being a body), he impressively explained by using the example of an actor. Actors (as human performers) are split between the role-playing subject and the subject acting in a role (cf. Soeffner 2001, 167).

Plessner argued that the performing actor emblematises 'condition humana' because theatrical plays constantly show the difference between the represented self and the self-representing self. Whilst acting, the actor steps out of himself to represent a character through the material of his own existence. In this way, the actor shows the duplication and the distance of the self (Fischer-Lichte et al. 2004), as well as showing how imitation can be used in everyday life to solidify sociological theories.

Techniques of imitation can best be described using Goffman's terminology. Therefore, I utilise Goffman's thoughts about communication to develop my own research.[21] Goffman's description of the social world is found in his book *The Presentation of Self in Everyday Life* (1959). In this, he argues that members of society intentionally present the image or self (self-presentation) they wish to present to the world. He backs this up by developing a dramatical approach to sociology. He believed that most people want other people to hold a desirable impression of them, and so they create such an impression, in a technique that could be called impression management. To advance his approach, Goffman used theatrical terminology to describe this phenomenon. According to Goffman, in order to present a good self-image and to avoid being embarrassed, people use techniques like 'surface acting'. Goffman described this to as a face-to-face inter-action, similar to theatrical performances, in which the management of the self is the most important task. Goffman then described how actors also actively manage their self-impressions by using the medium of the stage. Through the use of certain images (expressional repertoire), actors can 'play' social roles, and personal characteristics or items (such as clothing, posture, age, gesture, facial expressions) are also part of this play.

Goffman further distinguished between behaviour (that shows which role an actor wants to play) and appearance (which informs us about their social status). In the theatre, actors perform in the front stage, while actors relax, prepare, and sometimes play their social role backstage. Goffman stated that in situations of co-presence, social actors sometimes communicate more than they intend to. When relating this back to general society, Goffman coined the terms 'expressions given' and 'expressions given off', which describe situations in everyday life. In a theatrical frame, Goffman used the term Dual I, which described the difference between the I of the stage performer and the I of the staged figure/character (Goffman 1974). This should be understood as an embodied mask, wherein the role (a specialised function) and the subject (a personal identity) are not the same. If we apply this, scientific experts on stage at Science Slams must be seen as the dual combination of the representing self and the represented self. In this situation, the scientist's staged identity is probably not the same as the scientist's own personal identity. The scientific persona is a collective identity that does not necessarily correspond with that of an individual, but shapes the aspirations, characteristics, lifestyles, and even physical abilities of a group that is committed to this identity (Daston 2003).

Since Science Slam presenters can personally control the scientific persona they present on stage, it is interesting to see how they stage themselves. Science

Slam organisers proclaim that Science Slams offer an insight into the backstage of science. If we apply Goffman's ideas to the Science Slam genre, front stage is where scientists publicly carry out their performance and backstage is where the performance is prepared. Similar to commercial recordings of orchestra rehearsals (Goffman 1974, 145), it could be argued that in Science Slam events, backstage as well as front stage is staged. Science Slam performers present informal scientific knowledge on the front stage. Additionally, though, some elements of backstage life are expected to become part of the front stage in Science Slams, for example, the private personality of the scientist, or their everyday research.

Communication and the Study of Lectures

The thoughts and ideas about contemporary public communication and science communication that we have explored show that it is difficult to leave behind the narrative of the cave allegory and carry out a performative or communicative turn. The transfer of communication is viewed as a success if it fulfils 'the achieved transfer of information from one party to another' (Bucchi 2008, 66). In arguments about the incapability of social performances in keeping up with ideal conditions, known best in Chomsky's (1981) work (for example, his ideas on formal grammar), social performances are presented as unsuccessful re-enactments of something pure. In interpretive perspectives in sociology (and probably in fields like communication studies, media studies, and cultural studies also), the historically established opposition between ideal content and unsatisfactory form seems a little outdated. Contemporary developments in cultural studies and the rise of new terms such as 'performativity' (Butler 2002; Knoblauch 2017, 2008b) and the 'theatric' (Fischer-Lichte et al. 2004) raise questions about the existing strict dualism between content and form (or, subject and object). Since the 1980s, according to Fischer-Lichte, new approaches in the field of cultural studies have attempted to rewrite the traditional European model in which the Western world was presented as a text- and monument-oriented culture, while non-Western culture was subordinated.[22] Such approaches try to show that contemporary Western culture is not merely based in text, but is also present in performative processes (Butler 1998), or in '[p]ractices which form the objects of which they speak' (Foucault 1972, 49). Recent developments in the sociology of knowledge point towards the communicative construction (Knoblauch 2013, 2017) of society. Both of these changes use the connection between content and form, and knowledge and knower as a productive basis for society.

Next, I will present an alternative concept of communication. In his 1955 Harvard lecture *How to do things with words*, J.L. Austin highlighted how statements (language) could be viewed as performative. Austin's (1975) work was part of a wider movement that was growing at the time, which was focused on a performative turn in theory. This led to the introduction of two new ideas in sociology in the 1950s. The first was the dramaturgy of social life, and the second was the performativity of everyday activity. Hymes (1971) and Butler (1998) later expanded on this idea of a performative framework.[23] The work of all these authors, plus

Warner (2002) raise the question: to what extent is the act of naming a performative process? Arguably, we must clarify whether the announcement (the act of naming) of a particularly creative and entertaining scientist who is to take place in a Science Slam results in a performative process (e.g., one of making 'creative/entertaining scientists').

We previously explored Goffman's belief that the printed text of a presentation was the most important aspect and should therefore be the focus of attention. For him, the content of a talk was often in conflict with the way the content was staged. If the content was boring, this allowed for the attention of the audience to slip and for their attention to focus instead on the way the content was performed. In the struggle for attention between text and performance, the text often had a difficult starting position.

> Observe, I am not saying that audiences regularly do become involved in the speaker's subject matter, only that they handle whatever they do become involved in so as not to openly embarrass the understanding that it's the text they are involved in. In fact, there is truth in saying that audiences become involved in spite of the text, not because of it; they skip along, dipping in and out of following the lecture's argument, waiting for the special effects which actually capture them, and topple them momentarily into what is being said—which special effects I need not specify but had better produce.
>
> (Goffman 1981, 166)

Goffman claimed that giving a scientific presentation was a paradoxical and risky move, from which exertion was produced by risking the content. This is why he was more interested in the different roles that a speaker could have. He distinguished between the animator, who animates the text, the principal, who believes in the text, and the speaking machine, who presents the text. In this context, the presented self and the social performance not only have communicative functions but are also important for the individual. Although Goffman was a pioneer in some ways, his definition of communication may seem a little outdated now. He defined communication as 'the transmission of information by means of configurations of language signs' (Goffman in Ytreberg 2009, 7).[24] Although Goffman is a reflection of his time (mid-20th century, United States), his definition should not be misunderstood as old school. He was interested in the tension that arose between intentional and non-intentional communication (expressivity) because he believed that 'when an individual appears before others his actions will influence the definition of the situation that they come to have' (Goffman 1959, 6). His arguments about academic lectures in *Forms of Talk* (1981) can be understood as a belief that shadow theatre is always part of communication. In this sense, one could say that distractions can be ignored but cannot be wished away. This is especially true if lecturers 'come equipped with bodies' (ibid., 183). By this, Goffman meant that if a lecture contained visual and audio effects, the audience could easily be distracted from the speech. Unintentional expressions that are given off, on the other hand, are always part of communicative processes, because communication is not a perfect transmission from sender to receiver, but rather situated communicative action (Knoblauch 2017). In the 1980s, Goffman offered more of an insight into his views on the study of lectures when he argued that a lecture should be seen as more than just 'text transmission' (Goffman 1981, 186). Goffman claimed that the element of noise was one possible reason why people would want to attend a lecture (the music of interaction).

> For what is noise from the perspective of the text as such can be the music of the interaction—the very source of the auditors' satisfaction in the occasion, the very difference between reading a lecture at home and attending one.
>
> (Goffman 1981, 186)

There were, according to Goffman, other compelling reasons for wanting to attend a lecture. Firstly, attending a lecture allowed an audience access to the self of the speaker (the issue of access). The audience would gain information about the author or authors, and they would also have the chance to build a one-way relationship with him or her. Secondly, the audience could become part of a celebrative occasion (ritual character), where the uniqueness of the present is central. Further elements such as contextualising devices, responsiveness to local situations, and sentences that belong entirely to a particular setting (spoken register) could foster a feeling that they, the audience, were attending a good talk. A lecture was also, according to Goffman, a social event that was not faceless or anonymous. It could also act as a warrant: 'through evident scholarship and fluent delivery the speaker–author demonstrates that such claims to authority as his office, reputation, and auspices imply are warranted' (ibid.). Lectures give access to a topic, through a situated, vivid presentation. The content is designed for the event, and the self is designed for the context. In order to believe that the science is driven by trustworthy agents, the author of the knowledge became very important, within public science communication at least. This is especially true in differentiated societies. If we contextualise Goffman's thoughts, it can be argued that what science communication analysts call 'noise' and 'shadow' seem to entice people into going to science communication events, such as the Science Slam.

Communication and Communicative Constructivism

What Goffman called noise embodied communication was specially addressed in the empirically grounded communicative constructivism work of Knoblauch (Knoblauch 2013, 2017). Using Berger and Luckmann's work on social constructivism (1967) and his own empirical research on knowledge communication as a base (Knoblauch 2007, 2013), Knoblauch came up with a new theoretical framework he called 'communicative constructivism' (2013, 2017). Communicative constructivism was fundamentally based on critique and was therefore a revision of prior social constructivist theories. The prefix *social* was replaced with the prefix *communicative* so the focus in communicative constructivism was on the actual modus of *social construction* in an empirical sense. In this, action was regarded as communicative.

> Because social action requires a form of objectivation allowing us to coordinate our conduct with that of others in a way that makes sense to others, it is, in fact, communicative action.
>
> (Knoblauch 2013, 162 [transl. MH])

In Knoblauch's theory of communicative constructivism, the term objectivation meant something slightly different from the definition given by Berger and Luckmann. Objectivation, for Knoblauch, was the third area in the triadic structure between subject and others. Knoblauch further differentiated between objectivation (Objektivationen) and objectivisation (Objektivierung) (Knoblauch 2017, 163ff). Objectivations were seen as bodily processes that took place in the presence of communicative action, while objectivisations were perceived as if independent from the body. With reference to Barad's (2003) 'agential cut', Knoblauch also explained how causes of action could be perceived as if cut off from actions themselves. This reifying cut made 'the third' become an objectivation. What Barad called 'thingification', meanwhile, was called objectivation in communicative constructivism. Within this framework, communicative constructivism allowed for a very broad term of communication, which explicitly included Goffman's 'signs' and 'signs given off'. The body, its performance and 'cut off' objectivations are central to this approach, which I largely follow.[25] My research additionally seeks to reveal more about communicative construction (Knoblauch 2013), particularly within developing events in science communication. The developments in science communication are reflected in the accentuation of the role of informatisation, digital media, and performance in the face-to-face communication of scientific knowledge.

Defining Science Communication

According to some scholars, a distinction should be made between science communication and public science (Robertson-von Trotha and Morcillo 2018; Bauernschmidt 2018). Science communication is the 'top-down', institutionalised communication of science at universities, while public science is more of a voluntary 'dialogue-oriented' communication, following in the tradition of Humboldt. Knoblauch saw science communication not as a perfect transmission of scientific information from sender to receiver, but as a situated communicative action (Knoblauch 2017). My use of the term science communication in this book refers to the communicative action of scientists, based on a theoretical concept of communication and on the understanding of science that I have developed. More specifically, when scholars or scientists talk to each other or to a non-academic with knowledge they refer to as their 'scientific expertise' (insofar as this reference is essential to the form of communication), I call this science communication. When analysing science communication in the communicative genre of the Science Slam, meanwhile, I look at the validity of the scientific knowledge. I also refer to the communication (and partly the intersubjective validation) of scientific knowledge. For something to be labelled scientific in communicative construction, it should be in the form of either an embodied subject, their objectivation or other (triadic perspective). In my project, I focus particularly on communication in action. Therefore, what I call science communication here is beyond (and partly prior to) the publication and reception of ready-made science via scientific papers. It refers instead to face-to-face situations (what I call 'in action').

Discussion and Conclusion

Through the description of European science communication programmes, sociological reflections on science communication, and empirical data from the genre of the Science Slam, in this chapter I have shown how difficult it is to move beyond Plato's allegory of the cave and onto a deficit model of public science communication. The concern, shared by many scientists, that the only thing we find in science communication is 'shadow theatre' or 'slapstick', as in the Science Slam, is still relevant. This fear is also present in new communicative genres like the Science Slam. In this chapter I have further argued that events like the Science Slam should not automatically be seen as indicative of cultural decay. Through the use of framework from the sociology of knowledge, I proposed that we should not automatically assume that our initial judgements about knowledge are true. In a time of post-truth politics, the sociology of knowledge must concern itself with whatever passes for 'knowledge' in a society, despite the potential invalidity of said knowledge.

In the second half of the chapter, I developed an understanding and definition of communication. I also explored how the search for performative processes, and the analysis of the staging of the self, have been of major interest to sociologists since the 1950s, and I suggested that one should use Goffman and Knoblauch's concepts of 'communication' for empirical studies. Next, I argued that unintentional factors will always be part of communicative processes. I suggested that access to a speaker's self, and celebrative occasions are major incentives for people to want to attend academic and public lectures. I argue that we need to study the performativity of science communication in more depth. As we have seen, some researchers have already looked into this and have highlighted the multimodality of communication (Knoblauch 2007; Kiesow 2014; Tuma 2012). Such approaches, especially those from the interpretive paradigm focused on interaction, are helpful for further research on embodied science communication processes. Within STS, there is an interest in the status and role of communicating scientists. However, this rarely includes other interactions and material forms that accompany and help to establish a scientific body of knowledge in public science communication. Therefore, I have developed a heuristic, which allows one to study knowledge and science communication in a broader sense.

Notes

1 'The social power of scientific knowledge and rhetoric is huge, because the power of believing in the truth creates the appearance of truth' (Bourdieu 1988, 71).

2 Inscription 'refers to all the types of transformations through which an entity becomes materialised into a sign, an archive, a document, a piece of paper, a trace' (Latour 1999, 306); 'The Great Divide can be broken down into many small, unexpected and practical sets of skills to produce images, and to read and write about them. . . . writing and imaging cannot by themselves explain the changes in our scientific societies' (Latour 1990, 4). Consequently, for Latour, non-human entities are as stabilisers of actor networks.

3 'STS imported the phrase "social construction" from Peter Berger and Thomas Luckmann's, *The Social Construction of Reality*'. Ian Hacking later correctly pointed out that a lot of things were said to be socially constructed (cf. Hacking 1999, 1).

4 The first form was the micro-oriented consensus constructivism (or constitutive constructionism) which was rooted in the contract theory of Hobbes and Durkheim. The second form was called ideological constructivism (or conflict constructivism), because it aimed to understand why people disagreed so much about things that appear to be common realities. This perspective was concerned with the question of how the social location of something shapes the individuals' perception. Abbott saw the theoretical roots of this form of constructivism in the work of Mannheim, Goldmann, and Hauser and most importantly Marx. The third form he named structuralist constructivism. This had its roots in the work of Levi-Straus and Saussure.

5 Lynch proposed instead that the reader study ontology as an empirical topic.

6 In recent developments in communicative constructivism in the sociology of knowledge, the term 'objectivation' has evolved to mean something different from the definition given by Berger and Luckmann. In the new way, objectivation is seen to be the third area in the trio of subject and others. Knoblauch (2017, 163) also differentiated between objectivation (Objektivationen) and objectivisation (Objektivierung). Objectivations are viewed to be bodily processes, done in the presence of communicative action, while objectivisations are perceived to be as if independent from the body.

7 The works of Evans-Prichard (cultural anthropology) and Mary Hesse (relativist philosophy) have also been mentioned often.

8 The sociology of science also relied on the critical thoughts of Peter Winch on Merton's work, and used these to develop a method that took scientific enterprise into account.

9 Haraway provocatively stated that the only people that ended up actually believing in the ideological doctrines of science were non-scientists (Haraway 1988, 576).

10 'If it is no longer clear that scientists and technologists have special access to the truth, why should their advice be specially valued?' (Collins and Evans 2002, 3).

11 Sheila Jasanoff disliked Collins' narrow reduction of STS's writings on relativism. 'What concerns me here is the persistently reductive reading of sources that the argument exemplifies. For more than a generation now, work that C&E rather casually group together as "Wave Two" has formed part of a project of social theorizing that goes far beyond the relativism which the authors focus on. To say that work as diverse as the one Brian Wynne, Michael Callon, Bruno Latour, and for that matter my own, represents Wave Two's preoccupation with relativism is to seriously misread this work' (Jasanoff 2003, 391).

12 Schütz famously argued that as long as humans are born from their mothers, the alter ego experience is genetically and constitutionally prior to ego experience (cf. Schütz 2003, 115).

13 'Gegenständlichkeit als solche . . . für ein entfremdetes, dem menschlichen Wesen, dem Selbstbewußtsein nicht entsprechendes Verhältnis des Menschen. Die Wiedeeraneignung des als fremd, unter der Bestimmung der Entfremdung erzeugten

gegenständlichen Wesens des Menschen, hat also nicht nur die Bedeutung, die Entfremdung, sondern die Gegenständlichkeit aufzuheben, d.h. also der Mensch gilt als ein
nicht-gegenständliches, spiritualistisches Wesen'. (Marx/Engels Werke 40, 575).'All
estrangement [alienation] of the human essence is therefore nothing but estrangement
of self-consciousness' (MEW 40, 575).

14 While current British, American, and French researchers are fighting against a certain
type of social constructivism, in Germany there are still serious issues with radical
constructivism. For this reason, social constructivism not only has to be aware of international developments in STS, but also has to continue the ongoing debate with radical
constructivism.

15 Hannah Arendt's thoughts were published in the *New Yorker*, and they caused a great
amount of indignation. She was accused of mixing up offenders and victims, trivialis-
ing and normalising Nazi crimes, and falling for a spectacle of irresponsibility.

16 Berger and Luckmann also acknowledged the special power that reification had as
legitimation: 'Reifications imply that man is capable of forgetting his own authorship
of the human world, and further that the dialectic between men, the producer, and his
products is lost to consciousness. The reified world is, by definition, a dehumanized
world. It is experienced by man as a strange facticity' (Berger and Luckmann 1967,
89).

17 'Objectivations are every cultural product (no matter if it is music or painting, milk,
cars or a parakeet) including the schemes that classify them' (Knoblauch 2013, 29
[translated by MH]).

18 Translated by Miira B. Hill from: accessed May 19, 2015, http://scienceblogs.de/
bloodnacid/2015/03/19/quo-vadis-science-slam/. This example was described as 'very
German' by participants of a science communication workshop in Copenhagen (organised by Maja Horst and Sarah Davies).

19 'I know that it may seem a bit stiff and patronising, but eventually organisers must ask
the question whether the Science Slam should (also) have an educational responsibility (of which I am personally convinced it should) or if it is allowed to just offer pure
entertainment, and whether it is in their interest that the Science Slam is discredited
by too much comedy and boulevard and too little original research'. Source: accessed
May 19, 2015, http://scienceblogs.de/bloodnacid/2015/03/19/quo-vadis-science-slam/.

20 This is similar to the ethnomethodological indifference in which researchers are
encouraged to systematically deny the idea of advancement in knowledge. Instead,
social concerns are reconstructed from social situations and from the member's perspective (Garfinkel 1967; Lynch 1988).

21 I also take into appreciation recent thoughts in 'communicative constructivism' (Keller, Knoblauch, and Reichertz 2013) which strongly tie the sociology of knowledge
with materiality and empirical research.

22 The Western tradition (which was described as text-oriented, while non-Western culture was described as body-oriented) seemed to reach an end. Culture was no longer
described in monuments, but rather in performative processes (Fischer-Lichte 2011,
2001).

23 Butler viewed performativity as a mode of identity production and fixation, while discourse was seen as a source of that productivity. She emphasised that discourse produced the effect that it labelled.

24 This quote is from Goffman's unpublished dissertation: Goffman, Erving, "Communication Conduct in an Island Community" (Unpublished PhD thesis, Department of
Sociology, University of Chicago, 1953).

25 Knoblauch (2017, 177) recognised the important role that the Leibkörper had in the
communication process and with it senses, affects, and subjectivity.

4 Materials and Methods

My empirical analysis is made up of a mixture of different types of data from the Science Slam field. During the course of my sociological research, which has a strong focus on ethnography, I was allowed to become a member of the Science Slam field (to a certain degree). This allowed me the opportunity to attend Science Slams as an affiliate of the Science Slam community, not just as a normal attendee. Thanks to this opportunity, it was possible for me to approach my research from an emic perspective. My work, aside from touching on conventional ethnography, has a theoretical focus. Essentially, I investigate how, and why, communicative actions develop within the complex and uncertain relationship between science and the public. Additionally, I examine how communicative actions have an impact on the legitimacy and quality of scientific work. By using a combination of methods, I wish to offer several perspectives. Research on focused ethnography (Knoblauch 2001) was particularly helpful in gaining a deeper insight into the Science Slam genre. I used a variety of different types of data in order to answer the 'how' and 'why' questions posed. In my research, I not only talked to organisers of Science Slams, I also analysed successful Science Slam presentations (first, as a participant and second, by analysing video recordings). To understand the context, reconstructions and the interpretations within Science Slams, I also analysed discourse found in documents from the field, such as newspaper articles about the events and information on Science Slam websites. In this chapter, I will specifically explain my methods, and my research strategy within this context.

Qualitative and Explorative Research Design

The methodology of my analysis attempts to capture the transformation which takes place in science communication processes. My choice of method is loosely based on the constructivist statement that 'reality is built on an active-constructive manufacturing process and therefore cannot be studied through a passive-receptive process of illustration such as statistical data' (Flick 2005, 21). Consequently, my theoretical and methodological framework is qualitative, as per the logic of interpretive sociology (*Verstehende Soziologie*). Following in the work of Schütz, Berger, Luckmann, and Knoblauch, I think that the Science

DOI: 10.4324/9781003172635-4

Slam can be understood best by analysing the first order reality-constructions and communicative actions of Science Slam participants. In this analysis, a combination of qualitative methods is useful in order to understand what the Science Slam is all about. My general aim was to use a combination of qualitative methods, as is recommended for focused ethnography (Knoblauch 2001). Several methods were used to collect different kinds of material. In a similar way to the principle of Grounded Theory, my data had to align with other data in a triadic formulation. Within verbal data, visual data, and observer data, the relationship between subjective knowledge (legitimations), the relation to others (performance, interaction), and the objectivated world (socio-material arrangement, language, and body) were of central interest. Some modes of data were only suitable for a certain type of interest (e.g., video analysis was helpful for studying the interactional relation between people). Through the use of these different methods, I aimed to identify and describe features of the communicative genre of the Science Slam. When I found evidence that proved a certain perspective in one source of data, I compared this to what I found in other sources.

My approach was based on explorative research design. Throughout almost all of my research, it was unclear which people, events, and activities would be at the centre of my investigations (Merkens 2009, 295). The cases that were finally identified as relevant were not fixed in a theoretically deductive way but resulted instead from the research process. I had to sift through many case studies in the course of my investigation in order to find the right examples, and to identify the right kind of data. This research style was partly based on Grounded Theory. Unlike Grounded Theory, however, my research process was also based on clear theoretical assumptions about communicative action and genre. By doing so, I gained some ethnographical knowledge at the beginning of my research (at first, I did not know which of the participants were successful or who were powerful organisers in the field). My research process design was, therefore, circular. It rotated between interviews, video analysis, and short excursions to the field and all three were closely related.

Focused Ethnography and Access to the Field

Unlike in classical ethnography, where long excursions to the field are the norm, in focused ethnography (Knoblauch 2001) shorter lengths of time in the field are preferred. In order to understand the everyday life and perspectives of Science Slam participants, I chose to use ethnography and participatory observation methods. As recommended for anyone performing qualitative research, I intended to keep an open mind. Therefore, in the beginning, I clarified my own expectations and knowledge, and avoided reading too much scientific literature about the subject. Instead, I attended and participated in several different science communication events, such as Performing Science in Gießen, Lecture Performances in the HAU theatre, TED Talks in Berlin, Science Showoff in London, and Science Cabaret in New York.

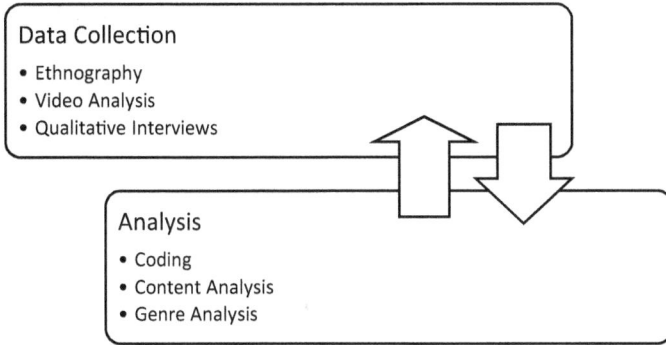

Figure 4.1 Overview of My Circular Research Process Design

When I attended Science Slams, I tried to capture the event's individual spirit and to understand its social dynamics. I found that the ethnographic process in which researchers take part in social situations was the best way to study the field (Knoblauch 2003). With this in mind, I attended Science Slams as an ordinary visitor and took note of my first impressions. I then contacted the respective organisers of the events and asked them if I could study the communication processes in more detail. I was offered backstage access to many Science Slams and I was also allowed to join in with gatherings that took place before and after the events. Due to the fact that there is very little research in the genre of the Science Slam, most people in the field were very open and happy for me to study them. Indeed, all the organisers that I contacted were very friendly, and more importantly, open to my questions. Many of the event organisers held small ritual get-togethers with food for the slammers before the presentations began and then also arranged for beers after the event. I was often invited to partake in these gatherings. It was not only the organisers who were open to my investigations, the participants were very welcoming too. It is important to state that I did not, at any point, expect to be able to participate in the Science Slam events as an affiliate; I was merely lucky I received such generous treatment.

During my experiences in the field, there was only one occasion where I was dismissed from the backstage of an event. The organiser later explained to me that she had asked me to leave because she was stressed and needed a bit of time with the slammers in order to regroup. This particular organiser was very concerned with protecting backstage and making it a safe place for the slammers. When she expelled me from backstage, she addressed me as if I were a member of the press. This was, however, the only instance where a Science Slam associate communicated to me that they felt uncomfortable with my presence. This one case of mistrust proved to be an exception, and all of my other experiences were full of compassionate and hospitable people. It seemed that most were happy about my research project and even happier to be part of my investigations (in fact, several

Science Slammers even contacted me after their events because they wanted to be interviewed). During this process, one thing that aided me was the use of the snowball method (Merkens 2009, 293), in which well-known people in the field introduced me to other relevant events and people. After conducting interviews with some important people in the Science Slam field (one being the founder of the Science Slam), I then gained access to other events, like the annual meeting of Science Slam organisers.

Genre Analysis

As stated, when we touched on theoretical framework, knowledge about communication is often conventionalised (institutionalised). One goal in my research was, therefore, to understand the patterns of institutionalised communication. The study of communicative genres focuses on the ways in which actors coordinate meaningful communication over time (Günthner and Knoblauch 1994). Genre analysis is a theoretical and methodological orientation, which is studied alongside the institutionalised communication processes of Science Slams. The study of communicative genres puts (1) internal structures (language characteristics, guiding motifs, topoi, and media), (2) situated realisation (performance in relation to an audience including gestures, facial expressions, interaction rituals), and (3) external structures (institutional context, social structures) into a central position. In my research, internal structure and situated realisation were both analysed through video material. Contrastingly, external structure was reconstructed by interviews and contextual investigations. Through the use of genre theory, I analysed communicative actions and looked for underlying communication problems.

Usually, genre analysis starts with recording a situation (1). In my research, I recorded natural situations using a video camera. The second step of the genre analysis process is the preparation of transcripts from relevant situations (to the research question) (2). I started with very rough transcripts of situations in order to gain an impression of what was going on in those situations. At first, I avoided reproducing fine grain transcripts because I wanted to identify relevant cases during the research process, not through genre analysis. The analysis of some presentations (even if they were not relevant) served as the basis for my subject-founded approach. Initially, these presentations were analysed according to generic features, and I developed hypotheses about rules, purpose, form, content, time, place, and participants. The third step of genre analysis involves hermeneutically analysing transcripts (3). As soon as I had identified interesting and relevant transcripts, I then prepared more detailed transcripts, based on an everyday understanding of words and sentences. In this step, the knowledge from my ethnographic experiences was especially helpful. The final step of both genre analysis and my research involved the analysis of the relevant sequences in data sessions (using Conversations Analysis) (4). Through all of these steps, I developed structural models for both internal and external structures, as well as for the performed dimension of the Science Slam genre.

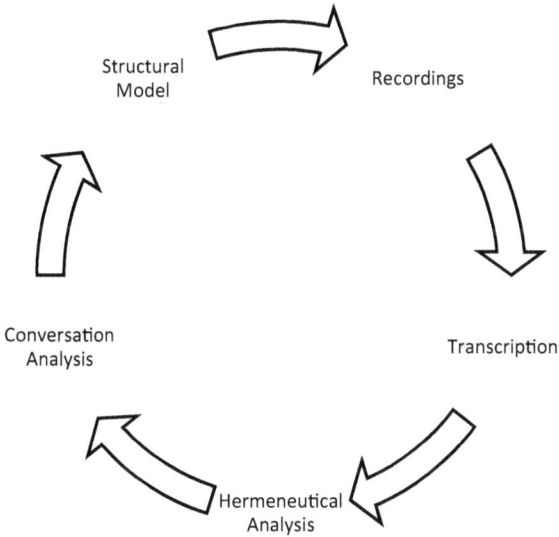

Figure 4.2 Overview of Genre Analysis

Video Analysis

A suitable method for dealing with data on a performative level is video analysis. If the production of objectivations is crucial for communicative action and the communicative construction of reality, then video analysis is a helpful tool with which to analyse them (Knoblauch 2013, 29). As previously mentioned, communicative actions are essential for the communicative construction of reality (ibid.). Observing genres also involves observing the manifestation of communicative patterns in action and performance. I am predominately interested in the interactions between the speaker and the audience. To understand the performativity of these interactions, I have used what has been called the 'microscope of interaction studies' (Büscher 2005). With video analysis, micro-analytical observations of a social situation are possible, while focused ethnography becomes more focused and slides into videography. The advantage of observing participants through video data is that it allows one to make intersubjective observations intelligible. In audio-visual data, language, gestures, facial expressions, postures, formations, roles, clothing, prosody, and noise from socio-spatial arrangements can all be studied (Tuma, Schnettler, and Knoblauch 2013, 5). In this way, the video camera is the 'interaction microscope' (ibid., 6), analogous to a microscope in a scientific laboratory. Acoustic and visual effects in interactions, in particular, can be described very well through video analysis. Although some senses (such as smell) cannot be captured by video, most other senses can be examined very closely. In my first material evaluation, I focused on capturing the generic characteristics of

Science Slam presentations. This first analysis of the material revealed a particular typology of the genre-specific performance.

Sociology distinguishes between natural social situations and artificial situations (like a movie set). The Science Slam is a natural social situation, because the video camera is a normal part of the event. All of the videotaped situations would have taken place regardless of my presence as a researcher and the slam would have been videotaped either way. Thus, the usual issue of being unable to predict participant reactions to the camera in videography is not a major issue in my audio-visual data. The reactions to the camera are natural because, in Science Slams, people expect to be videotaped.

Like genre analysis, video analysis focuses on sequences that are relevant to the research question. Transcripts of these sequences are then prepared and analysed in order to develop a deeper understanding of them. Since my ideas were still vague at the beginning of my explorative research, I randomly selected a few presentations myself and analysed them. I also analysed transcripts in group sessions in order to come up with a shared understanding of the sequences. Additionally, I participated in several data sessions with other researchers, and I discussed various sequences and transcripts with scientists in my direct academic environment (for example, in a Data Session at the Institute of Sociology at the TU Berlin). Furthermore, I collaborated with scientists in the wider interdisciplinary field (for example, at a Video Analysis Workshop in Tübingen), and with sociologists and videographers from other universities (for example, in a Video Analysis Workshop in Berlin).

The Specific Problems of Data Collection and Sampling

Videography is a process which enables the interpretive video analysis of social situations (Tuma, Schnettler, and Knoblauch 2013, 7). In the videography process, the production of one's own video material, which is then analysed, is preferable. The usual process involves the researcher searching for relevant social situations and then focusing on these situations with the camera. These situations are also usually shot and recorded from several different angles or points of view.

I recorded many Science Slams with my video camera, but this resulted in two problems. First, there was an issue with visibility. In my first video recording, I focused on both the speaker and the audience. Although the speaker was visible in the video, the audience were in the dark, due to the lighting in the room. At first, I thought this was a problem. However, I later realised that the audience's visibility might not be as important as I had initially thought, mainly because the speakers did not seem to see much of their audience and so the audience was unlikely to affect the speaker's performance. Further examination of the interactional sequences on my videotape confirmed that speakers were indeed struggling to see the audience. Whenever speakers were interested in the audience's response to a question, they had to lift their hands up to shield their eyes from the blinding lights in order to be able to see the audience. One could say that the interactional connection between speakers and their audiences in this context can best be described as an acoustic connection, rather than a visible one. An additional issue concerning visibility was the fact that Science Slam organisers often claimed the best recording positions and they also had better, higher resolution cameras. In many cases, I compared my own recordings of events with the official recordings and I recognised that, from a video analytical perspective, I would see more if I used the official recordings.

The second problem I had was caused by my sampling strategy. Unfortunately, by the time I decided to focus solely on the most successful Science Slams, I had already seen all of the slams at least once, but I did not have videotapes of them all, so was unable to analyse them. Due to this issue, I decided to work instead with video data from the internet, which I complemented with my own recordings, and my general ethnographic knowledge. Broadly speaking, video analysis should only be used if one has a vast amount of knowledge about their chosen field and the social situation of that field. It was only due to my previous ethnographic research that I was able to interpret the data I found on YouTube and on other internet sources to its full extent. Thus, I was able to combine my 'ethnographic knowledge' from my work in the field with the data I had found. In my case, this technique worked quite well because, as we will see, Science Slams as an institutionalised genre have many specific characteristics. As such, I was able to put the situations I saw into perspective and then identify three things: what roles people played, what else was known about the setting, and whether situations were typical or atypical. In general, I preferred to include video recordings in my analysis, at least in those cases where I had additional ethnographic knowledge.

My primary source of data was audio-visual material from public science presentation events. The first aim of the sampling strategy was to understand what usually goes on in Science Slams. After my first few field trips, the focus of my observation became the development of a general understanding of observed actions. After attending several Science Slams, I learnt that there was a group of science slammers who toured around Germany and were successful in several situated settings. I became increasingly interested in highly prestigious slams, especially if they were viewed as particularly good or innovative by science communicators. This became one of my major interests, and it eventually led me to analyse several successful presentations myself. In order to compile some features of successful slams I looked at three things in the video analysis. Firstly, I looked at 10 of the most successful presentations at situated Science Slam events. Situated Science Slam performances where my main interest. Secondly, I looked at 10 of the most successful (or most watched) Science Slams on YouTube and, finally, I looked at the four winning presentations from the German Science Slam Championship.[1] There was, inevitably, some cross over and instead of having 24 slams to examine, the total number of slams examined was 19, because some slams were duplicated across these three formats. My analysis of these 19 events allowed me to identify several key features of successful Science Slams. In a similar way to the concept of theoretical sampling in Grounded Theory, my first hypotheses were based on my research question. I was aware that Science Slam presentations could vary according to a number of factors: the different actors, the background of the Science slammers, and the organisational context. Therefore, I tried to come up with a broad perspective which would cover all of these conditions. As we will see when I cover the external structure of the Science Slam later, it is clear that Science Slams also often have different financial dependencies, institutional contexts, particular social milieus, and rules of communication, which vary vastly from one to the next.

Qualitative Interviews

I used qualitative interviewing as a third method for collecting data (Kelle and Erzberger 2005, 303). In order to practice interviewing and to get an impression of the field, I started by conducting short interviews (around ten minutes in length) with people from Science Slam audiences. I then started interviewing well-known individuals involved in Science Slams, like the presenters and organisers of events. I prepared an interview template to use, but I always tried to develop new questions based on statements given by the interviewees. My questions varied greatly but one important question was focused on understanding the features of the Science Slam. Another important part of the process was to find out what problems the establishment of the Science Slam had caused. Further interview topics included questions about the occupational history of the interviewees, their history with Science Slams, their motivations and purpose, as well as examples of Science Slam standards, questions about coaching practices, opinions about university lectures, and discussions over who was a favourable slammer. Questions about the setting, the atmosphere, and the audience were assessed too.

The interviews took between 1 and 1.5 hours. Some of them were recorded in public spaces, prior to or after a Science Slam event, but many were done via Skype, due to distance. I did, however, ensure that I only interviewed people via Skype if I had already met them face to face. I promised my interviewees that I would keep their answers anonymous. As such, the only answers that will be identified are those relating to the establishment of the Science Slam. From these interviews, and from various newspaper articles, I gained an impression of what the genre was like historically. Of course, *ex post* descriptions can be inaccurate and newspaper articles are typically exaggerated, but since there was no other data to draw from, I had to rely on these two sources for information.

Luckily, most of my interviews had a relaxed and trusting atmosphere. Only one of my face-to-face interviews had a strange social dynamic.[2]

I found that after I had interviewed some important slammers, I was then able, through word of mouth, to interview other relevant people, and so on (the snowball method). Although the snowball method generally gives clumped samples, it allowed me to find out information about people, events, and activities in Science Slams that I wouldn't have discovered without using it. It also allowed me to find suitable cases and interesting areas for my study. In some of the later interviews I even tested concepts and results that I had developed.

Through my interview data, I learnt a lot about the establishment of the Science Slam, and about the outer structure of the field. Since there is almost no scientific literature on this phenomenon, I had to generate relevant contextual knowledge myself, and in doing so I learnt even more about Science Slams.

Coding

Coding is the application of categories and their interrelations. The term 'code' is also a general concept for the conceptualising of data. Therefore, a code is a result of an analysis. In a similar way to the coding paradigm of Anselm Strauss, interactions, conditions, actors, and policies were described (Hildenbrand 2009, 36) and coded in my data. Similarly, I formed hypotheses based on my material. In my data, in general, I searched for an indication as to why, and how, particular communicative actions came about in the relationship between science and the public. I assessed what was viewed as legitimate, and high-quality, science in this context. I worked with the programme MAXQDA in order to code the material. I also analysed 14 German Science Slam event websites and looked in particular at the similarities and differences in the way they presented Science Slams, as well as their own self-representation. I then used MAXQDA to highlight frequently occurring motifs and references in the Science Slam. Lastly, I coded the qualitative interviews, the video descriptions, and my field protocols.

Data Corpus

My empirical work is based on an analysis of a huge number of different types of data. It seems helpful to give an overview of this data. So, the list of data is as follows: over ten hours' worth of material from interviews with Science Slam organisers (10), documented experiences in the field of ethnography (20), an analysis of Science Slam websites' content (14), and an analysis of the characteristics that successful slammers possess (19). To supplement this data, I also video analysed the ten most successful Science Slam presentations and held short interviews with audience members from some German Science Slams (six). I supported this data by using quotes from some newspaper articles, although this was rare.

I did not use all of the data I had found in my research. In addition to the data that I did analyse and use, I also collected data from various other sources. Among other things, this allowed me to compare and contrast the different types of data, before deciding which data types integrated best. The other data I collected came from: interviews with slammers from Berlin (eight), interviews with slammers from Hamburg (two), interviews with slammers from Cologne (seven), interviews with performers from the UK (six), and interviews with organisers from the UK (one). In addition, I took videos of two events in the UK, where I was also a participatory observer (two), I held interviews with Science Slam organisers in New York (two) and videotaped an event in New York. I also participated in the German event 'Performing Science' in Giessen in order to learn about Lecture Performances.

Transcription

I tried to find the right style of transcript and to make sure it matched my data sources. My interviews mainly focused on why those involved wanted to take part in slams. I was not really interested in the way that people said things, but more in the contents of what they said. This was good because the interviews were often roughly transcribed and so there was little attention placed on prosody, pauses, and overlapping. Another point worth mentioning is that many of the interviews were translated from German into English, so language translation mistakes were made. On the whole, I tried to correct these because I wanted to ensure that the content was represented right. However, the reader should bear in mind that some quotes from the interviews are not congruent with the German version.

With the video analysis, I was interested in understanding the communicative actions, as well as the content, in more detail. Compared to the standards of conversation analysis (Sacks, Schegloff, and Jefferson 1974), my video transcripts were rather rough. Due to the fact that the events and interviews were originally in German, the original transcripts are therefore in German. As stated before, I tried to generate adequate translations but, as the reader will see, the prosody and the emphasis on certain words differs between German and English. The point of the English translation was not to make the interviews linguistically better, but to try and explain clearly what was going on in German. I think the transcriptions are adequate for the level of analysis attempted here, and that my rough style of transcribing is helpful for understanding the communicative actions of Science Slams. I used the following conventions for the transcripts:

But al-	disruption of a statement
a=a	fast connection
(.), (_),(1.0)	short and long discontinuations of speech, with an approximate break. Time of the break in seconds
.	terminative
,	continuative
?	appeal
(?)	unintelligible word
(hope?)	uncertain transcription
[...]	non-transcribed speech
[laughs]	non-phonetic communication from speaker or audience
[unintell]	longer unintelligible speech
this is GOOD	special emphasis on a word or syllable
a:nd a::nd	extended or stretched pronunciation of a syllable
underlined	emphasis
me	part of speech that is accompanied by a non-linguistic act will be italicised and described under the transcript row
(points to his chest)	

Figure 4.3 Overview of Conventions of Transcription

Methods Revised

Feminists like Haraway argue that ideas such as embodiment, partiality, and localisation should be grounding practices for knowledge claims (objectivity). If I could read just my empirical access, I would push my method into the direction of feminist ethnography. The ethnographic concept of Abu-Lughod (1990) for example is an argument for a partial perspective. She emphasises that feminist ethnography is by no means a question of irreversible relativism, since this would simply be a reflection of total objectivity. Denying the relevance of location, embodiment, a partial perspective, and total relativism are seen as the same 'God trick' that scientists try to use whenever they obscure their own position. Thus, feminist ethnography could be seen as an opposition to the godlike gaze of the unmarked, white, male subject. A method of 'ethnography of the particular', which Abu-Lughod proposes, is grounded prior to fieldwork with the focus on stories about specific individuals in time and space. This perspective leads to self-critical reflections.

My ethnography is that of someone who graduated from Western universities, researched in Western institutes, and has been involved in Western discourses. I am a tall slim woman with African American and mid-European roots. My family does not come from a typical middle-class or upper-class background. My father studied anthropology; he was the first in his family. My father was black, but he was not accepted in the USA, neither by the blacks nor by the whites. They called him 'the grey'. My mother grew up in rather poor circumstances in Germany. She had to experience a lot of discrimination in her school years because of poverty. She and her younger sister were left behind by her mother at an early age with her father, who was traumatised by war. She started working at the age of 14. She trained as a nurse. Later, in the time of the New Age movement, she went to India and became a nun. She was a nun for 7 years before she met my father in the USA. My father lived as a transgender for a while before he met my mother. My father felt he was a woman in love with women. I am the eldest of four siblings. My parents got divorced. My mother later was a yoga teacher. My father did not get his educational qualification recognised in Germany. He worked as doorman, taxi driver, and later geriatric nurse. Nothing was normal in this family. It felt like we were the ones in between.

Notes

1   I collected this data on July 14, 2014. Since then, this information might have changed because the successful Science Slam videos on YouTube are variable. The most popular Science Slam video on YouTube is '*Gut with Charm*' by Giulia Enders. It has more than one million views (1,088,800 at the time of writing). The second most popular video on YouTube is '*Women's Theory*' by Robert Idels (625,912). He studies Mathematical Economics. The third most popular slam on YouTube is '*Entropy: About Cooling Towers and About the Irreversibility of Things*' (227,218) by Martin Buchholz. His research comes from the field of Applied Science and he studies Thermodynamics.
2   I felt as if my interviewee wanted everything he said to be perfect, so it felt a little bit like an examination. I therefore excluded this interview from my study because I felt that there was nothing natural about it, and that all his information was just a response to the unpleasant interview situation.

5 The Science Slam as Communicative Innovation

How Was the Science Slam Established?

In this chapter, I will discuss to what extent we can talk of the Science Slam as an innovation in scientific communication. In the first section of this chapter, I will explore the important features of the sociological and social constructivist perspectives on innovation. My focus on subjective knowledge, the relations of people, and the objectivated world (socio-material arrangements, body, and language) shows that all three levels have an impact on the institutionalisation of a new genre. I will further remind the reader of the long tradition of public science communication that I have already outlined in this project.

We can observe the influence on the Science Slam of numerous institutionalised ways of talking about science, and of sharing science with the public, which are part of a long history of similar scientific communication events. In the 17th and 18th centuries, for example, phenomena like the public experiment aimed to make science tangible through entertainment, and to legitimise scientific research. Aesthetic experiences, eyewitness accounts, and a controlled observation of new knowledge were all practiced publicly long before the Science Slam came into existence. In the following pages, we will discuss why these principal functions have once again become relevant, and also explore what new features have emerged in modern science communication events.

Due to the fact that it is difficult to identify a single origin of the innovation process, I suggest that we stop worrying about the birth of innovation as a concept. Instead, I will argue that Science Slam participants frequently talk about novelty, and want to communicate in innovative ways, but that they also wish to reproduce established structures as well. This chapter will further clarify, from the perspective of those involved in slams, what it means to organise or participate in Science Slams. A sociological study should always include a field reconstruction from a subjective point of view, and for this reason, I have included selective interview excerpts from participants, wherein they speak about their own experiences at Science Slams.

Introduction

Historians hold contrasting views about how, and why, the interest in innovation, invention, and 'the new' emerged.[1] What many historians agree upon, however, is

DOI: 10.4324/9781003172635-5

that the term 'innovation' had to go through a process of transformation in order to acquire a positive valence. After a period in which innovation was prohibited (pre 19th century), it became instrumental and more prominent in the 19th and 20th centuries onwards. Yet, it was not until the 20th century that the term 'innovation' became common in Germany. This may have been due to the increasing emphasis on the economy that was prevalent at the time, or to the popularity of Schumpeter's economic cycle theory, which suggested that innovation was the main reason for development in society.

In the present day, innovation has become an 'object of veneration and cult worship' (Godin 2014b, 8) and a value in itself. Nevertheless, sociological reflections about modernity and Western culture have occasionally been critical about the presence of teleological progress in economy and society. Theories that view some societies or cultures as more or less advanced, or which categorise cultures according to stages of higher and lower civilisation, have also been criticised. In addition, innovation-oriented societies have been universally criticised for their will to shape the future which, in turn, leads to a neglect of the past, which many view as problematic. In such future-oriented societies, the labelling of something as 'new' is often linked to forgetting about achieved knowledge. In the book *Dialectic of Enlightenment* (first published in 1944), the classic critics of modernity, Adorno and Horkheimer, described how belief in progress and innovation was a major problem for society. Adorno and Horkheimer argued that the cultural industry creates the impression of being innovative but does not offer any new ideas or theories. In the mid-20th century, the exclusion of anything 'new' or 'novel' was seen as an indication that a culture was infecting everything with 'sameness'.

> Unending sameness also governs the relationship to the past. What is new in the phase of mass culture compared to that of late liberalism is the exclusion of the new. The machine is rotating on the spot. While it already determines consumption, it rejects anything untried as risk. In film, any manuscript which is not reassuringly based on a bestseller is viewed with mistrust. That is why there is incessant talk of ideas, novelty and surprises, of what is both totally familiar and has never existed before. Tempo and dynamism are paramount. Nothing is allowed to stay as it was, everything must be endlessly in motion.
>
> (Adorno and Horkheimer 2002, 20)

Later, sociologists came to a similar conclusion when they argued that, with the domination of mass media, the focus on new forms of communication in society guided the way towards a paradoxical leaning towards novelty. As social communication processes were increasingly located in mass media, a greater focus on novelty developed, while novelties themselves continuously faded away (Luhmann 1996). In contrast to Adorno and Horkheimer, the French scholar Miège (1989) argued that cultural industries are perfectly capable of producing new aesthetic innovations. If one agrees with this perspective, it can be argued that the sheer repetition of symbolic arrangements is not satisfactory for members of society.

It has been suggested that in experience-oriented societies (Schulze 1992), the pleasure of consuming adventures is based on the ongoing alternation between embodiments and receptions of embodiments (Hutter 2015a, 2015b). In aesthetic capitalism (Reckwitz 2012), the economy focuses on consumers who are searching for consumable experiences (Hutter 2011). The call for subjective experiences and surprise creates a demand for a permanent novelty of form (Hutter 2015a). Creative industries and the experience economy focus on selling experience, and design novelty as an end in itself. Thus, aesthetic experiences have become of intrinsic value. Aesthetic capitalism depends on the economic practice of permanently offering new products to consumers in order to stimulate aesthetic experiences (ibid.). Despite the fact that the Science Slam is not very profitable, it is arguably part of the experience economy because it offers consumers experience-based goods.

Classical Concepts of Innovation

In economics, the term 'innovation' is widely associated with Schumpeter's (1883–1950) work on economic cycle theory (*Konjunkturzyklentheorie*). Schumpeter suggested that innovations are crucial for society to develop, but that there are both negative and positive consequences of describing innovation as 'creative destruction' (Schumpeter 2000).[2] It has been argued that Schumpeter supported the 'Great Man Theory', because he believed that inventors produce ideas, and entrepreneurs act on them. Indeed, Schumpeter ([1934] 2011) viewed entrepreneurs as individuals 'who exploit market opportunity through technical activity and/or in the organisational innovation'. Entrepreneurial activity was understood by Schumpeter to be the implementation of a combination of materials and forces in order to produce new things, or the same things by a different method (ibid., 51). In this sense, Schumpeter's views on innovation were widely associated with the economic body. Traditional economic approaches to innovation focused on studying processes that would end with the establishment of commercialised products. In this sense, the inventor and first entrepreneur of the Science Slam, Alexander Deppert, can be viewed as having created a new type of science communication. Nevertheless, since the Science Slam is generally not profitable, Schumpeter would not view the Science Slam as innovation.

In the 1930s, several new sociological theories on innovation, with a focus on the social, emerged. These theories were not just interested in the establishment of commercialised products, but in the diffusion of social inventions into wider society. The first sociologists to touch on innovation in a sociological sense, at this time, were Ogburn (1886–1959) and his colleague Gilfillan (1889–1987).[3] Ogburn was interested in social change and wanted to understand why social change occurs and also why, in certain circumstances, change is resisted. Ogburn (1922) argued that inventions depend upon cultural factors, preparations, and cultural needs. For Ogburn, inventions were the result of a material and technological environment that emerged through an evolutionary process (Godin 2010, 14). This contrasts with Schumpeter, who believed that great inventors were responsible for creation. Though Ogburn's theory was problematic in parts,[4] it was an important early work that accurately described the different stages of sociological innovation without using economic parameters or the 'Great Man Theory'. Ogburn criticised the 'Great Man Theory' by shifting his attention to cultural path dependencies and societal needs, thereby affirming that cultural factors have an impact. Ogburn's research is helpful because it endorses the argument that material and technological factors (like the invention of PowerPoint), and cultural needs (the need to explain scientific research to the public) made the Science Slam possible.

Innovation from the Perspective of Social Constructivism

In contrast to some well-known sociological approaches to innovation studies (OwenMacnaghten, and Stilgoe 2012; Zapf 1989),[5] Knorr-Cetina (2002) defined scientific innovation in a socio-material way and proposed that innovations go through two stages of success. Firstly, according to Knorr-Cetina, innovations have to be successfully constructed in a laboratory and, secondly, they have to be accepted by other scientists (Knorr-Cetina 2002, 123). Along similar lines, Braun-Thürmann (2005, 6) defined innovations as 'material and symbolic artefacts that are observed as novel and as better' (than what came before). According to Braun-Thürmann, every innovation is the product of practices and structures of society that have been acknowledged by people. This definition allows for the belief in progress of society; it does not have a pro-innovation bias, because what is labelled as novel or better does not actually need to be novel or better. Definitions such as these allow one to describe a society that believes in empty innovations as cornucopian.[6] In Suchman's (2011, 15) words, innovation can end up being an 'articulation that calls out a difference from whatever is referred to as the thing that came before'. Thereby, innovation can be seen as merely prolonging already well-established social and technical arrangements.

By borrowing concepts from the sociology of knowledge and through observing, and talking to, innovation researchers, Knorr-Cetina (2002) concluded that ideas that lead to material and interpersonal success solve problems by using analogy. Innovation is, according to Knorr-Cetina, a transfer of established solutions to a new area, rather than the result of open problem-solving. The conservative nature of innovation comes from this link to successful problem-solving strategies in the past. Innovative ideas mobilise existing and established knowledge (ibid., 112), while innovation and acceptance are temporary stabilisers in the process of knowledge construction, which is fundamentally a social process. In order for the development of the new to be successful, commitments, conventions, expectations, successful problem-solving strategies from the past and, finally, situated experiences with materiality are all key.[7] I agree with Knorr-Cetina's definition, and therefore understand innovative public science communication as a consequence of successful Science Slam performances in various situated settings. From a sociological perspective, there is no innovation without social acknowledgement. In my work, I question what legitimate, scientific, high quality knowledge is in this context by looking at universally acknowledged successful Science Slam presentations. My work is not, primarily at least, concerned with the constitution of innovation (there is good reason to suggest that there is no single origin of innovation), but more concerned with questioning what makes a Science Slam performance 'good' or 'better', and what judgements are used to come to such a conclusion. I mostly align with the work of Hutter et al., Braun-Thürmann, and Knorr-Cetina. Nonetheless, it seems necessary to outline all the potential origins of innovation in my theoretical framework of the sociology of knowledge.

The social constructivist perspective defines innovation as 'the communicative construction of something as new' (Knoblauch 2014, 8), and this includes knowledge-based difference.[8] Crucially, this perspective acknowledges the role of knowledge and communication in the creation of novelty (ibid.) and, as I will show, this is helpful if one looks at novelty in a communicative genre.

For Schumpeter (2011), innovation was established through entrepreneurial activity and by people thinking up new combinations. Schumpeter believed in the 'Great Man Theory' and therefore argued that an inventor produces ideas, and an entrepreneur disseminates them. For Knoblauch (2011), meanwhile, innovation or novelty could come in two forms. Inspired by Schütz, Knoblauch described how important fantasy was as a catalyst for imagining the new, and he concluded that the imaginary (ibid.) holds great significance.[9] Although fantasy is subjective, and therefore not universally accessible, it can be visualised, expressed, verbalised and, in this way, become accessible (Herbrik 2011). In fantasy, knowledge and action are not separate, but coupled entities, so fantasy can be a guiding point for action, and a trigger for new combinations (social constructivism talks about social structures, cultural needs, and material constellations). The second source of novelty that Knoblauch described was situated creativity and situated action. While fantasy relates to the imagined and happens only in the mind, situatedness is highly dependent on perceived socio-material constellations (the physical) and action (the experimental).[10]

Studies on innovation are typically based on the essentialist view that we can access innovation. By taking a social constructivist standpoint, I will use Schutz's postulate of adequacy (cf. McLain 1981) and argue that an understanding of innovation must be grounded in everyday life. I am interested in finding out what is acknowledged as innovation in what context. However, the innovation must be something materially realised in order to assess how social context has had an impact on it. Like Braun-Thürmann, I see innovation as an entirely social category, which means that some things are more 'novel' or 'better' than other things. I am not suggesting that less industrialised countries should have Science Slams as well, or that the Science Slam is more novel or better than other forms of science communication. I argue, instead, that there is a specific social context in which people think that the Science Slam is a great, novel way to communicate scientific knowledge. I am simply interested in discovering what passes for innovation in public science communication. By taking the social constructivist perspective towards innovative science communication, I am interested in the intersubjectivity of innovation and in the institutionalised stock of knowledge.

> From the actor's point of view—both producer and the related 'recipients' or users—the new then is characterized by a meaning that is different from their typifications and their existing knowledge. As these typifications of differ-ence must be relational, this knowledge needs to be shared in such a way as to make the difference explicit or objectified.
>
> (Knoblauch 2014, 7)

Genres of Communication

Thomas Luckmann recognised that we have to empirically study communica-tive genres in order to learn more about the social construction of reality. The subject of genre has been analysed by various disciplines including sociology, anthropology, linguistics, media studies, and communication studies. Within sociology, genre analysis is closely connected to the communicative paradigm, which has been used to understand the overall design of communication. The communicative paradigm is very closely related to Goffman's theoretical con-cept of frame analysis (1974), which describes how one situation can become a model for another. One goal of the social constructivist research programme is to understand the patterns of institutionalised communication. Communicative pat-terns are typically able to free people from underlying communicative problems related to action (Luckmann and Knoblauch 2000; Knoblauch and Raab 2001). Nevertheless, genre should not be misunderstood as functionalist. In a similar way to frame analysis, where the defining question is 'what is going on here?', genre analysis involves analysing communicative processes that are perceived by par-ticipants as closed. Through genre, people can coordinate their action more easily and communicate via organised pathways. For this reason, the study of commu-nicative genre predominately focuses on the ways in which actors can coordi-nate meaningful communication over time (Günthner and Knoblauch 1994). All knowledge, including knowledge about communication, is conventionalised and institutionalised.

> All orderly formal public communication of one to many relies upon such conventions, and the extent to which precision of thought is communicated is a function of the precision of the procedures, spoken or written.
>
> (Mead and Byers 1968, 5)

Günthner and Knoblauch (2007) described communicative genres as a frame or orientation of people's actions. According to Günthner and Knoblauch, all communicative genres should have (1) an internal structure (which includes characteristics such as language, guiding motifs, topoi, and media), (2) situ-ated realisation (a performance with an audience, in which gestures, facial

expressions, and interaction rituals are common), and (3) an external structure (which is defined as institutional context and social structure). All of these
should involve realisation in action.[11] It is important to note that the analysis
of an external structure highlights the interrelatedness of communication and
its institutional setting and social structure. Communicative genres as a whole,
then, can be described as communication forms that are guided by certain
motives, themes, and topics, and have observable physical and linguistic features or, in other words, 'typified communicative actions characterized by similar substances and form and taken in response to recurrent situations' (Yates and
Orlikowski 1992, 299). An important feature of both current society and communicative genres is reflexivity. In society, individuals structure their actions
reflexively, and this coordination extends to communicative genres. There is a
reflexive form of self-monitoring wherein communicative genres have rules and
follow complex action patterns. Since members of society can coordinate their
actions easily, without much negotiation, the internalisation of these rules and
patterns eases communication processes.

Novelty in a Communicative Genre: The Endless Becoming

Many theories on genre give the impression that it is a rigid object of study.
Yet, genre is dynamic; it represents conventionalised, but flexible expectations
about how communicative procedures should be organised (Günthner and Knoblauch 2007, 65). Genre is constantly recontextualised, updated, and one genre
can become a model for another. To enact genre means to change and revise it at
the same time, but an original activity in a genre remains important in order to
understand a new one. Additionally, the guidelines and performative realisation
of genres are in a reflexive relationship. In this way, every previous performance
in a genre can potentially be a course of action. The communicative arrangement, and all objectivations in a genre (like the layout of PowerPoint slides) can
become a model for another, while character staging scientific personas or translation practices can have an effect on the performative acts of others. People's
actions are orientated towards meaning and this meaning is based on their experiences in everyday life. So, one could argue that genres are social institutions
that 'both shape and are shaped by individuals' communicative actions' (Yates
and Orlikowski 1992, 300). This process happens in the midst of the triadic
structure between ego, alter-ego, and objectivation. When studying the actuali-
sation of communicative genres in situations of interaction, an analysis of genre
leads to an endless becoming of communicative genres. Genre is the vehicle
and outcome of communicative action (ibid., 302).[12] Genre forms a frame that is
interpreted and altered with every action of its members, so genre is subject to
ongoing change and modified expectations. When a new genre is established, a
new shared stock of knowledge about communication appears, along with new
ways to interact. Taking in all of this, one could conclude that Science Slams are
a form of typified communicative action, which occur in response to recurring

situations. In this chapter, I will use the idea of habitual communication in order to detect novelty.

> While members typically reinforce established genres, [they do so] both deliberately and inadvertently. When changes to established genres are repeatedly enacted and become widely adopted within the community, genre variants or even new genres may emerge, either alongside existing genres or to replace those that have lost currency.
>
> (Yates and Orlikowski 1994, 6)

So, we can call something a 'new genre' when the communicative expectations of participants change. In their research, Yates and Orlikowski (2007) considered the subject of derivative genres that arise in relation to established genres. They argued that older genres influence newer ones because participants of derivative genres build their expectations in relation to older ones. Thus, Yates and Orlikowski questioned when it is legitimate to say that a new genre has developed.

> In practice, it is impossible to define an exact point. The above definition of genres as socially recognized types of communicative action suggests that variants of communicative actions are still recognizable as instances of the old genre, while a new genre can be said to have emerged when a new conjunction of form and purpose becomes recognized by its community as different from the old. Such recognition may be explicitly articulated within the community or be implicit in members practice.
>
> (Yates and Orlikowski 1994, 6)

In a similar way to the division between semantics and pragmatics, the separation between the 'explicitly articulated' and the 'implicitly practised' highlights the diverse ways in which a new genre can appear. If we accept Yates and Orlikowski's definition, we can only label a genre as 'new' when a community recognises a new conjunction of form and purpose (as different from the old). If this is true, new derivative genres are always historical, and always relative to society.

When a new genre is established, a new shared stock of knowledge about communication, as well as new ways of interacting, become accessible. In my analysis, I rely on various different sources of data in order to specify or describe the Science Slam. It is possible that genres change not because of their structure, but rather the structure varies with the form of communication.[13] As we established earlier, slam participants distinguish between scientific and non-scientific knowledge, so knowledge about science and knowledge about genre are reflexive categories. Thus, my analysis of genre highlights both the knowledge and the situated practice of slam participants.

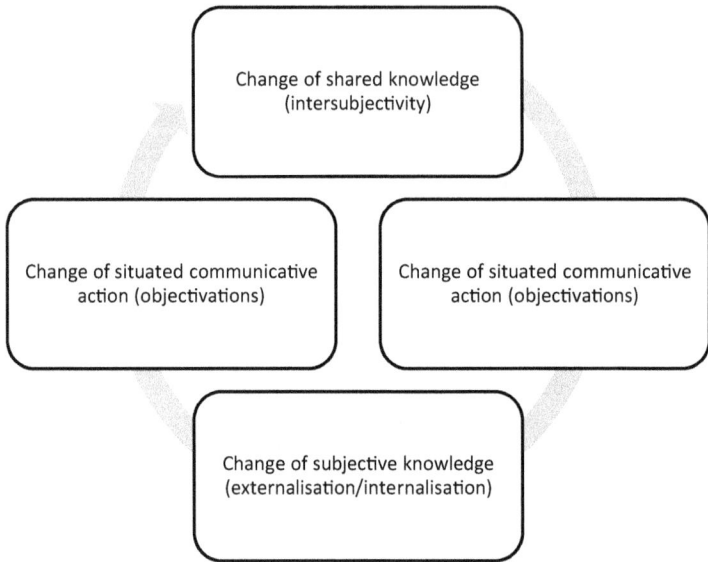

Figure 5.1 Overview of Knowledge and Situation Communicative Action

Historical Prototype—the Public Experiment

The Science Slam is a new conjunction of form and purpose. However, like all new genres, the Science Slam's roots lie in the past (Mead and Byers 1968, 4). These roots reach back to the 18th century, where there was an interest in creating a form through which scientists could communicate with the public and inform them about scientific research. This resulted in the so-called public experiment, an important historical ancestor of the Science Slam. In the public experiment, scientists would perform scientific demonstrations in front of the monarchy and patrons. The aim of the public experiment was to prove scientific concepts or theories and establish a new intellectual order, as well as making science tangible and entertaining for the audience. It was a practice of 'letting spectators see for themselves' (Smith 2009, 451). Famous public experiments demonstrated concepts such as electricity, physics, and chemical processes. The public experiment made the natural sciences both accessible and perceptible for the bourgeois and can be viewed as a prototype of the Science Slam (Shapin and Schaffer 1985; Krifka 2000b).

Although the public experiment is of particular interest, it is not the only ancestor of the Science Slam. As we explored earlier, the communication of scientific ideas in salons or coffeehouses was common in the 1650s. This practice continued through to the 18th and 19th centuries. Although they varied in form, what these ancestors of the Science Slam had in common is that they all involved face-to-face communication of scientific knowledge. Science Slams should be seen as the modern equivalent of such historical events.

As well as exploring the historical ancestors of the Science Slam, it is important to find out why, and how, new science communication genres are staged, which will be done using Goffman's concept of frame. Goffman argued that people use a large amount of energy in defining situations and overcoming insecurities, and to do so, they depend on processes of frame analysis and ritualisation (conventionalised acts) (Goffman 1974). Events such as the public experiment and the Science Slam all come with particular framings, and expectations about how the science explained in them should be communicated. As Smith (2009, 453) pointed out in his paper *Theatre of Use*, scientific demonstrations can be understood as a re-framing of scientific work.[14] Smith argued that modern public experiments (including Science Slams) are a presentable form of messy, private experiments. Thus, experimenters and scientists do not really 'perform' experiments, but rather create a showcase of their scientific knowledge.

To properly apply the concept of re-framing it is necessary to name not only historical prototypes, but also three of the most recent ancestors or prototypes of the Science Slam—small conferences, PowerPoint presentations, and the Poetry Slam—and to describe the development of these.

Recent Prototypes—Small Conferences, the Use of PowerPoint, and the Poetry Slam

In their work on 'small conferences' (or meetings), Mead and Byers (1968) discussed how new genres of communication have been received in relation to past and present social communicative requirements. According to Byers and Mead, there was a development in the 20th century, wherein traditional scholarly and political talks evolved into small conferences. After the Second World War, in particular, the small conference became a form of communication that contrasted with the typical, formal, and hierarchical style that had been common in organisations. The small conference allowed for informal and equal communication, gave each participant the right to speak, and was also spontaneous, non-linear, and simple. Although the genre was principally established to challenge traditional communicative arrangements, it also allowed new material arrangements to flourish. Traditional communicative arrangements have been described by Mead and Byers as such:

> The presence of hierarchy has been manifested in the construction of buildings, in the elevated pulpit, the raised lecture platform, the special seat for a cabinet minister, all giving elevated and visible status to those who spoke versus the permanently or temporarily humble of the auditors who were to be informed by those who spoke.
>
> (Mead and Byers 1968, 4)

Broadly speaking, the small conference involved small groups sitting around a table, at a specified place and for a specified period of time, in order to discuss a topic. There is no need for a linear script in small conferences. Instead, small conferences should be 'multisensory interchange with speech as the principal medium; attitude, shifts in attentions, gestures, and the types of expressiveness that cannot be adequately represented in print, play an important part' (ibid., 5). Although the small conference was a new form of communication in the 1950s, it had roots in the past as a social invention (Mead and Byers 1968, 3). The need for effective face-to-face communication in small groups originated, or at least had roots, in historical phenomena like small gatherings of humans in villages or tribes, and in scholarly communication. The interest in establishing a conference-based form of communication arose in the post-war context. Since small conferences are now well established across Western society, one might conclude that they were an innovation in communication at that time.

Similarly, the creators or pioneers of PowerPoint changed business communication in the 1980s, at a time when companies were starting to use visual devices (Tufte 2006; Knoblauch 2007). Today, PowerPoint (PPT) presentations are one of the most dominant forms of communication in business. According to estimates, around 30 million PPT presentations are made every day, while the software is installed on roughly 250 million computers (Günthner and Knoblauch 2007). Communication styles have been known to have been adopted across different contexts. This is certainly true of PowerPoint. It evolved from simply being a form of communication in business, to becoming dominant in society in general, and in science communication. PowerPoint has altered communication and performance among companies and scientists alike.

Knoblauch and Gunthner (ibid.) described the key characteristics of Power-Point and explained how PowerPoint has allowed many key speakers to expand their repertoires. These characteristics include checklists, paraphrasing items, the use of visual images, and the use of the body. Knoblauch argued that with the introduction of PowerPoint into talks, the triple combination of speaker, talk, and audience evolved into a four-way combination of speaker, talk, audience, and slides. According to Knoblauch, PowerPoint is not a one-sided method of communication. Instead, it is a mutual method of communication because it is dependent on the presence of a speaker and an audience. Either way, it has widely been accepted that knowledge has become more digital, a trend which has been reinforced by the practice of uploading slams to the internet and viewing them as a product (ibid., 19). Unlike Tufte (2006), who thought PowerPoint should be abandoned because it had caused a mental enfeeblement of society and blamed it for catastrophes like the crash of the Challenger spacecraft, Knoblauch argued that we should still study processes of communication like PowerPoint in order to learn about new developments.

A third contemporary prototype of the Science Slam is the Poetry Slam. The Poetry Slam was established during the 1980s in the United States in an attempt to democratise the business of literature (Westermayr 2004), and it spread to Germany in the 1990s. Petra Anders' (2008) studies on Poetry Slams indicate that Poetry Slam performances are often characterised by a connection to the daily news and daily activities. In the Poetry Slam, communication often includes interaction with the audience, while the rules include a focus on melody, on keeping it brief and on a certain play with alliteration. As we will later explore, Science Slam events are more based on visuals than Poetry Slams and they fit into a slightly different template. Yet, even if the Science Slam is typically described in relation to the Poetry Slam ('it's like the Poetry Slam, but with science'), Poetry Slams and Science Slams clearly distance themselves from each other. Science Slam organisers claim that poetry slammers reacted negatively to the Science Slam due to their belief that the Science Slam 'stole the format'. Poetry slammers, meanwhile, contend that the Science Slam is the bourgeois version of their original, anti-establishment event.

As many of the aforementioned works have demonstrated, genre varies according to the conventions of different socially situated groups. For this reason, I believe that we should analyse the Science Slam as a form of communicative innovation and explore how the genre's expectations and communicative actions have changed over time. I see the Science Slam as communicative innovation because it has changed traditional public science communication immensely and has had both situated and interpersonal success.

This project seeks to answer how contemporary challenges in communicating and legitimising science are handled in the Science Slam. In previous chapters, I have principally characterised science communication as the communication of scientific knowledge. More specifically, for me at least, science communication occurs when scholars or scientists talk to each other, or even to non-scientists, and refer to their scientific expertise. To come to this assessment, I have looked at the communication and intersubjective validation of knowledge in Science Slams. My understanding is that trust in institutions and society is key. In previous chapters, I have presented the theoretical basis for my research, and there are theoretical pre-assumptions in my argumentation that I will not bring into question. I prefer, instead, to closely investigate how systems of expectation about communicative change come about. I have principally argued that we should take a triadic perspective towards knowledge and science communication, which involves subjective knowledge, the relations between people, and the objectivated world. Therefore, these are the three relevant points of analysis to use when studying science communication and focusing on the social construction of reality.

The Development and Characteristics of the Science Slam

In this section, I will look at the history of the Science Slam, for which I rely on both interviews and newspaper articles. In a classic Schumpeterian sense, my narrative might seem to support the 'Great Man Theory', because it seems to necessitate the inclusion of both an inventor and an entrepreneur. However, my research shows that equivalent but different science-based slams have evolved in several countries in the West since the 1980s, which may indicate that the Science Slam does not necessarily support the 'Great Man Theory'. This chapter will show that such events have incrementally changed the institutionalised ways of communicating science and, also, that there is a corresponding tendency to frame something as 'new'. I wish to find out why, and how, communicative actions have arisen out of the relationship between science and the public, and how the scientific legitimacy and quality of these actions are evaluated in such contexts. To answer these questions, I will outline traditional communicative actions used by people, the objectivations typically used and why those involved in slams take part. A final important thing to note is that a genre must be understood as working under the conditions of permanent change (e.g., change in the expectations of participants and in situated performances).

The Invention of the Science Slam

In 2001, the psychologist (specifically, cognitive scientist) and poetry slammer, Alexander Dreppec (Dreppec is his real name, but he goes by Alexander Deppert so I will refer to him as Deppert from now on) wrote a dissertation in which he discussed the comprehensibility of scientific texts (*Verstehen und Verständlichkeit: Wissenschaftstexte und die Rolle themaspezifischen Vorwissens*. In English: *Understanding and comprehensibility: scientific texts and the role of theme specific prior knowledge*). As research for his dissertation, Deppert carried out two studies. The results of the first study revealed that readers with specific previous knowledge learn more from less coherent texts than they do from highly coherent texts. In the second study, he questioned to what degree 'geek-speak' or

Image 5.1 Alexander Deppert

technical jargon is part of English and German scientific texts. He discovered that German and English readers attribute a lower academic status to people who write in an incomprehensible way. Thus, he argued that there should be more effort placed on writing coherent scientific texts. In his spare time, Deppert took part in Poetry Slams and was known professionally as Alex Dreppec. He was awarded the Wilhelm Busch Prize for humorous poetry in 2004 and he also won several Poetry Slams in various German cities.

I interviewed Deppert, who thought it was obvious that he had invented the Science Slam principally because, according to him, the Science Slam combined both his research interest (psychology/science) and his favourite hobby (Poetry Slams).

> I had to have the idea, because I had researched comprehensibility and I performed in Poetry Slams.[15]
>
> (interview with Alexander Deppert, 2014)

Once we had established that Deppert had come up with the idea for the Science Slam, we explored his motivations for doing so. His initial motivation to create the Science Slam grew out of a dissatisfaction with traditional lectures and conferences. Broadly speaking, Deppert wanted to change academic lectures, improve interactional organisation at talks, and develop the scientific persona.

> When I came up with the idea of the Science Slam, I immediately thought that this could be an exciting idea, because . . . most academic lectures involve stiff, academic readings, with very little interaction with the audience, and the expectation that the 'holy poet' will clear his throat and then everyone must be quiet for one and a half hours, and at the end, everyone must clap. In Germany at least—although not in America—I felt that the atmosphere at lectures and conferences was like this. And then I thought, ok these are two really disparate ingredients.
>
> (interview with Alexander Deppert, 2014)

In 2004 Deppert submitted his Science Slam concept to the company he worked for at the time, Darmstadt Marketing, but it was not acted upon until two or three years later. Deppert's boss at the time supported his idea.[16] In order to make the Science Slam a reality, Deppert was given a little time off work and a one-off payment of €500. Despite this, most of the work done by Deppert to set up the Science Slam was voluntary and done in his spare time in the evenings and at weekends. Deppert's work did eventually pay off and Darmstadt Marketing was able to hold its first non-official Science Slam, which formed part of a programme called 'Wissen findet Stadt' (Knowledge Meets the City).

Before this event occurred, Deppert organised some initial tests, with friends, in a place called Teeladen. In this initial testing phase Deppert found that many people were willing to participate in slams, yet finding people who were also comfortable performing on stage was another matter, as was defining the content of slams. Despite these challenges, most of the first science slammers were colleagues or

Image 5.2 Alexander Röthemeyer

friends of Deppert. The first official Science Slam took place in September 2006 in a place called Kukicha (formerly Stoeferle-Halle'603 qm'). The slam was moderated by Alexander Deppert and Alexander (known as Axel) Röthemeyer. The seating was scattered around the room and dressing casually was encouraged, as was drinking beer on stage! Deppert told me that it was important to have objective signifiers that differentiated the Science Slam from traditional lectures at this early stage, thus indicating that the slam was an informal event.

> Difference signal. . . . Yes, a signal that this is not a conference, and not a stiff event. In the beginning . . . I asked them (the audience) not to come in suits. A friend of mine, who has become a relatively well-known specialist, Jens Hoffmann, won the first Science Slam. He brought his beer on stage. That made me happy. That was something that you would never do at a conference. And during his talk, he had his Hefeweizen next to him and every once in a while, he took a sip during his presentation. And that was so important to me, that you simply note that this is a slam and not a conference.
>
> (Alexander Deppert interview, 2014)

In the first recorded article about Science Slams, Deppert and Röthemeyer said,

> Often, students present excellent lectures, but unfortunately, they just do it once. At our event, students can recycle lectures.[17]

Deppert characterised the Science Slam in the beginning, as here.

> Yes, before I explained it with the Poetry Slam, but not all knew it, only a few knew. . . . And just as one would explain it today, it can be explained thus: it is a ten-minute oral presentation, in which presenters try to interest their audience, make their talk intelligible and try to establish a slam atmosphere.
>
> Furthermore, anything that makes the talk more vivid is acceptable. The value of the talk is not based on previous research, but on whether the lecture is thrilling.
>
> (interview with Alexander Deppert, 2014)

Deppert claimed that in the beginning, the Science Slam was closer to a traditional scientific talk, rather than a comedic one, although it was clearly more informal than traditional lectures.[18] Initially, the content of slams and the ways of presenting were not as strictly monitored as they are now. Today slams must conform to strict regulations (in fact, organisers meet once a year to debate the standards of the Science Slam). Additionally, knowledge presented in slams must now be original scientific research. Yet in the past, recycling research or lectures was common.

The first official slam was a huge success, so much so that the third slam (Science City Slam), which took place in the Stoeferle Halle as well, received media coverage. The slam was reported on by the Darmstädter Echo on 3rd December 2007, as the previous quotes show, and the headline for the article was '*Thrilling Lectures, Rhymed Distich*'.[19] In this article, Röthemeyer and Deppert were both named as founders of the Science Slam (later sources did not credit Röthemeyer). Interestingly, in the article, the journalist Marc Mandel (who is now the press spokesman of Science Slam, Darmstadt) pointed out that there had been no participants from the fields of natural sciences or engineering sciences. As a result, an advert was printed next to the article offering a free case of beer to the first three engineers to participate in a slam.

The Distribution of the Science Slam

Although the Science Slam initially began in Darmstadt, it was re-invented in a place called Braunschweig a few years later. This was initiated by Markus Weißkopf, who worked at the Haus der Wissenschaften (a local institution which aims to advance public science communication) as an event programmer. Prior to working at the Haus der Wissenschaften, Weißkopf worked for a consulting company and also acted as counsellor for the city of Konstanz. This is where the idea for the Science Slam came about. According to Weißkopf, a brain-storming session was held in Konstanz in 2005, 'and in this creative session, the idea to construct a Poetry Slam with science, and then call the whole thing the Science Slam' emerged. The event did not get implemented at this time and Weißkopf moved to Braunschweig to work at the Haus der Wissenschaften. Once settled in

Braunschweig, Weißkopf set about bringing his idea into practice but, after some investigating, he found out that a Science Slam had already taken place in Darmstadt, in 2006. He then decided to call Alexander Deppert.

> I called him and talked to him and he said he did it once and it is now currently on hold because he has other things to do, and then I re-established it in Braunschweig. It was funny actually; the ideas were so independent from each other. . . . He was the logical inventor, so to speak. . . . Yes, and it led to a run of slam events in Braunschweig.
>
> (interview with Markus Weißkopf, 2014)

> He called me, and I communicated the rules and other sorts of things.
>
> (interview with Alexander Deppert, 2014)

After all this had transpired, Weißkopf spoke about the Science Slam at a German science communication event (a conference on scientific journalism). Many people were intrigued by the idea and started presenting Science Slams in their own towns.

> Exactly, I met the organiser from the Braunschweig Science Slam, who worked at the Haus der Wissenschaften, Markus Weißkopf, at a science communication conference. Exactly. And he told me about it, and I was immediately totally fascinated.
>
> (interview with Julia Offe, 2014)

Julia Offe was one of the people who heard Weißkopf speak at the science communication event. She organised her first slam in May 2009 in Hamburg. Today, there are numerous Science Slam organisers, many of whom heard about the Science Slam either through the media or by attending science communication events.

> I read an article about a Science Slam in Spiegel. This was in summer 2009. It was about Julia Offe's event in Hamburg. She is probably happy when I say that. And so, I thought, that is a totally cool concept. Because it was a very funny article. And so, I thought to myself: why not do something new when I know that it works, so I gave it a try.
>
> (MS#44)

Many organisers attended other people's slams before organising their own event.

> In Germany, science communicators are quite well connected via Wissenschaft im Dialog and they host an event every year which is called Forum Wissenschaftskommunikation. This is a conference where science journalists, representatives of the DFG and the Stifterverband meet in order to talk

about science communication. . . . And at this Forum Wissenschaftskommunikation, in Dresden—it must have been 2012—there was a Science Slam
that we really liked. And we both recognized the genre from YouTube and
heard that it was proving to be successful in Germany. And then we wondered
why there was no Science Slam in our town. And then it was clear to us that
we have to do this ourselves.

(MK#45)

Setting up slams proved to be hard work and numerous organisers reported that
they had to work alone and take huge financial risks initially. One organiser
claimed that they did not talk about anything but their slam for six months. For a
slam to take place, many different factors had to come together. Most of the time,
slams only occurred because of a successful combination of highly skilled people,
a shared vision, finding an appropriate location, finding a willing moderator, and
being able to advertise.

So, I did it quite alone. What's assistance? I mean, I looked for a location
which I thought would fit. So, my innovation was ensuring that the event
took place in a bar or at a club and not just in the Haus der Wissenschaften
or such. So not just in the academic environment, but outside of it. . . . Then
I visited Poetry Slams in my town, something like five or six or seven. And
I asked the moderator I liked the most if she wanted to join me. And she said
yes immediately. Then I called the university's press office and asked whether
they could help me. And I had exactly the right person on the line. And he
asked, 'What are you planning to do?' And then I explained it briefly and he
said, 'Yes, yes, send me a banner in the size 100x200 pixels for our website's
homepage'. It was . . . that quick and a lot of people acknowledged that this
was a good thing.

(MK#46)

Yes, I did everything myself. I paid for it myself, or else I invested. I did
not make a profit, but a loss. The people who helped me I recruited from
the pub, because they were all passionate pubs-goers who watched soccer.
And the guy who shouted the loudest, funniest things in the pub, I asked
him to be moderator. He was hooked on it. There was also an amateur
musician in the pub. He was a little bit familiar with sound engineering
so I asked him whether he could help with the technology. And he also
recommended a location we could use. And then we went there together,
and the same guy knew the owner. Then we talked to him. He was thrilled.
So, one thing led to another. But, essentially, I did it all, in my spare time,
by myself.

(MS#44)

There are additional reasons why people have set up their own Science Slams.
Some organisers see their work as a necessary service because they think

there should be more communication platforms for scientists. Some argue that the responsibility for providing events like the Science Slam lies with the state.

> So sometimes I'm a little confused, because basically what we do is not our job, but it is the responsibility of the state, the country, the city, or the universities. Actually, we should really be asked and paid for it.
>
> (MS#47)

Arguably the Science Slam must be viewed as a pioneer of cultural improvement and support. One company, which had supported the genre financially for years, described its part in acting as a role model for other companies.

> But, in Germany, at some point, the government will no longer be able to spend so much money on culture, at least not taken from state money or from the federal states. They [Germans] will have to have the spirit of entrepreneurship. And there I see us as pioneers. Because we have never earned money from Science Slams. We paid for them for three years, that is what we made. And hopefully for a good reason—because we believe in the whole concept. There will surely be many more companies in the future, at least I hope, who will have this idea themselves, to start things for themselves and for their city, or for certain groups.
>
> (MS#47)

In addition to companies sponsoring slams, another way the Science Slam gained traction was through the practice of coaching, wherein slam organisers provided coaching services for those who wanted to participate in slams. This involved teaching potential slammers how to communicate correctly, how to appear comfortable on stage, and how to have appropriate body language, amongst other things. The goal was not to encourage a generic standard, but rather to develop slammers 'own language, and . . . own body language' (MS#47). It was all about loosening up communication.

> You can learn a lot. One, to keep it short and simple. And to step out of one's own habitat and one's own field, so to speak, and to communicate instead in the language of others.
>
> (MS#47)

The practice of coaching continues to this day, and scientists can learn a lot from being coached, for example, how to communicate in a comprehensible way. Those who have received coaching agree and say that the training also came in handy when they presented at conferences. In fact, organisers train both amateur scientists who want to participate in slams and professional scientists who make their living through science communication. A common practice in coaching sessions is to analyse videos of former Science Slams, in groups.

> I tell the participants that they should bring some YouTube videos and we watch them together. And then we analyse them and ask 'Okay, what was well done, what was bad?' One is then virtually in review mode.
>
> (MS#49)

In some cases, coaches ask slammers to do a test run of their proposed presentation in front of a group of organisers and moderators.

> We ask the slammer to prepare a presentation without our assistance and they bring their presentation to the coaching session. And when they are presenting, we do not hesitate to stop them and say, 'Wait a minute!' We also take notes and when everything is great, we say, 'Everything is great!' Then we describe the stage setup and explain where the computer will be and where their presentation will be. It makes no sense if a presenter turns around and clicks in the direction of the slides because nothing happens, it only happens if you click in the direction of the computer. Basics, right? We tell them a bit about what to wear on stage as well.
>
> (MS#44)

When a test run shows that a presentation is not good enough, organisers help improve the slam.

> And in extreme cases, we have had to create completely new presentations. We have said to them: 'You held this talk before at the university, right?' Now that's a positive reformulation of 'It was totally shitty and we need to make drastic changes!' You cannot say 'Shit!', but you can say 'Have you presented it at the university before?' No one comes to us without realizing they need help.
>
> (MS#44)

Other organisers do not offer coaching per se, but offer slammers advice, either face to face or over the telephone, to help them prepare for their slams.

> Many slammers need very intensive care. They are like, 'What do I put on the slides?! What version of PowerPoint do I have to use?!' And so, they are often very excited and ask, 'Can I wear a costume? What kind of microphone do we have?' However, very few have had substantive issues, so far.
>
> (MK#45)

Digital Distribution, Reflexive Production, and Creative Modification

We can only understand the social dissemination of the Science Slam in Germany if we look at the organisers' practice of uploading videos of slams to YouTube, which was established by the Haus der Wissenschaften in 2009 and has been common practice ever since. This practice means that there is a digital archive of the most successful slams online. On a separate but related note, some organisers watch YouTube videos of potential slammers before they invite them to be involved in their events. However, it is worth mentioning that although uploading videos to YouTube often leads to the slam or the slammer experiencing additional success online, it is not necessarily true the other way around. Success on YouTube does not automatically guarantee that slammers will succeed when attempting to book or take part in live performances.[20] Either way, the practice of digitally archiving slams online has led to organisers having to coach slammers on their behaviour, as we have already explored. With all this in mind, one could conclude that the Science Slam is both a situated event and a digital phenomenon.

> Others then, using completely different media, they appear more YouTubeish. I think that's pretty popular right now.
>
> (MS#48)

> On YouTube, the two/three-minute movies are the ones that are not very dense . . . on YouTube you cannot afford to have one minute of rest, because then everyone is gone. It's just that the attention spans are different, anyone can see that.
>
> (MS#48)

What we can take away from the practice of digitalising the Science Slam is how diverse informal learning and communication cultures can be, and how this learning can be mutually reinforced by both offline and online practices (Hill 2020). Slammers use various techniques to communicate their arguments; images, videos, and jokes are commonly used, as are scientific personas, references, and translation techniques. References and analogies, in particular, become viral and extend in time and space; for example, the reference to beer is often repeated multiple times in different slams. The translocation of social space (Knoblauch 2017, 341) is important in the practice of digitalising slams. When a slam is digitalised, it becomes transcendent[21] and mediatised, and the performance becomes part of a digital knowledge archive, so communicative actions are permanently present for those who have access to the internet (ibid.). According to Groys, archives are materialised memories of society (Groys 2004). This digital archive can be viewed by many people, including those who are planning to take part in slams themselves, which has proved useful in the past.

> It is good that YouTube is such a perfect coaching tool . . . because most people tell us, 'Before I created my slam, I first surfed for five hours on YouTube and looked at different presentations, to decide what is best for me'.
>
> (MK#45)

> One problem might be that the Science Slam creates uniformity, and then you realize that the new Slammer XY is a bit like another one from one or

two years ago. That's probably going to happen anyway because people are watching YouTube and learning from each other.

(MS#42)

As the quotes demonstrate, potential slammers watch YouTube videos of past slams in order to learn more about the genre. The critical analysis of past slams enables participants to develop their own style of presenting. By recognising mistakes made by previous slammers, participants can assess what they think works in a slam and what does not, which allows the Science Slam as a whole to develop, evolve, and grow. On the other hand, the use of this practice inevitably results in slammers using or borrowing aspects from other slams, a process I describe as 'reflexive production mode'. This technique of using features from other slams and reproducing them in a new context means that slammers can be compared to DJs who remix songs, or to social media users who share memes created by someone else (Fischer and Grünewald 2018). One example of this practice comes from a slam on Bioplasts by Simon McGowan, in which McGowan re-used elements from a previous slam on surgical implants by Peter Westerhoff.

In his slam, Westerhoff explains that he, and his team, 'build spare parts for surgeons', specifically implants. After explaining how successful they are at this, Westerhoff comically suggests that he and his colleagues do not 'Pimp My Ride', instead they 'Pimp My Implant'! This is a reference to the MTV show *Pimp My Ride*. Unfortunately, as Westerhoff explains, his boss does not look like the rapper XZIBIT (who hosts the show on MTV) but is different looking. Whilst saying this, he shows a picture of XZIBIT with a cross through it, next to a picture of his boss titled 'Big George Bergmann' (his boss's name). The next slide shows pictures of the rest of his team who he introduces using popularised American nicknames, for example, 'Iron Mike', 'Mister Slowhand', and 'The Brain'. Finally, he shows the logo of 'Charite Implants Custom', the company he works for, which is in the style of graffiti. Although Westerhoff is a specialised technician, he is clearly familiar with popular culture and manages to bridge the gap between the 'dry' sciences and 'cool' popular culture.

This reference to *Pimp My Ride* was later re-invented by Simon McGowan at the German Championship Science Slam in 2014, which took place in Berlin-Neukölln.[22] In his slam, Simon McGowan re-invents Westerhoff's reference to the show *Pimp My Ride* by showing a picture of himself on screen with the body of superman in front of a rainbow with a speech balloon that reads 'Pimp My Bioplast!' McGowan pronounces the word 'Bioplast' in an American accent. The writing used is not in the style of graffiti, as it was in Westerhoff's slam, but in regular font.

One might expect that when ideas are re-enacted or re-used, that questions concerning copyright would arise. However, in this case, at least, the re-invention of *Pimp My Ride* was not viewed as copying. It was viewed, instead, as the reproduction of something that already exists and the creative modification of it. After all, appropriation, re-enactment, and adaption of culture is common in both everyday life and in communication practices. Science Slam organisers recognise re-invention when it occurs and most are comfortable with it, although they would prefer slammers use their own original ideas, rather than borrowing too heavily from others.

> I think there must always be a chance that someone just does it a bit differently and that's why it might not be that useful if you watch too many YouTube videos.
>
> (MK#46)

> And then maybe I would say, ok, if somebody was really a copy, and I would not even say such a bad thing in public because you may be writing it down [laughs], then I would not like that person to be part of a slam event as much as I would like others to be. If something is original, I would say 'Wow, that's original, he really invented that himself' and I would like that.
>
> (MS#42)

Resistance

Organisers claim that most people are fascinated when they hear about the Science Slam for the first time. This creates the impression that the Science Slam is an important novelty and, thus, it is easy to create the framework needed for the slam to occur. Nevertheless, not everyone agrees that the Science Slam is novel or perfect, and it is important to explore the resistance to it, as well as the praise. Generally, organisers acknowledge two groups who dislike or pose a threat to the Science Slam. The group who presents the biggest threat to the Science Slam consists of 'old professors' and established scientists who prohibit their doctoral candidates from participating in slams. Their resistance to the Science Slam seems to be mainly focused around losing control.

> So, I've just met a Science Slammer. . ., who said she does not want to be in the press release, and she does not want to be filmed either because her professor has forbidden her to participate . . . Yes, and another slammer wanted to borrow something like a plastic model of a tadpole for her presentation, and her Professor did not give it to her. He said for such shit he would not lend her university equipment. . . . These professors do not know that communication is important; they do not realize that it is a great way for slammers to explain briefly . . . what they are studying for their PhD. Yes. And they fear that the science is trivialized.
>
> (MK#46)

The second group mostly consists of poetry slammers. The structure of the Poetry Slam was used as a template for the Science Slam, and the Poetry Slam model is also frequently used to explain how the Science Slam works. Many organisers have received complaints from poetry slammers, who claim that the Science Slam is a copy of the Poetry Slam.

> From the Poetry Slam scene, we are seen critically, of course, because we are the ones, so to speak, who have stolen the format. So . . . that is a complaint that comes often. Yet, I cannot quite understand why they complain, and would rather ignore it because I find that the contents are so seriously different that you cannot compare them. . . . But basically—in the Poetry Slam scene—they say that 'we stole the format' [laughs].
>
> (MS#47)

Some poetry slammers also view the Science Slam as the bourgeois version of their original, anti-establishment event and so, correspondingly, disregard the Science Slam.

> Yes, there are sceptical voices . . . who regard the Science Slam as the nerd amongst slams. Among poetry slammers there are some who claim . . . that they have no bourgeois foundation, and they live a life of rock 'n roll. And for them, Science Slams are where people who work in laboratories, or those who complete doctoral theses go. There are also scholarships available in the Science Slam. For them [poetry slammers], of course, this is a bit suspicious. So, I would express it like this: The Science Slam is the nerd amongst the slams. It is the most bourgeois. And that is certainly true if you look at the slammers' backgrounds.
>
> (MS#42)

Yet, the main problem faced by organisers remains to be the lack of recruits. Trying to find people who are willing to take part in slams is an ongoing, everyday task for organisers. Recruitment methods include advertising Science Slams in newsletters and calling students at universities to try to persuade them to participate. In fact, some organisers share the phone numbers of potential, or past, slammers with other organisers, to help them out. Several slammers told me that they find this practice annoying, and that they do not want organisers to assume that they are willing to participate in future slams, just because they have participated in a slam before.

> So, it's incredibly easy to find an audience. But to motivate participants to get naked on stage—in a figurative sense—is not easy. Not every scientist is a stage hog; many simply have the fear that they will be ridiculous or something. In the end, the reaction of the participants is consistently positive, and they often say, 'Well I did it and it's a whole new experience'. But early on, it is always a bit difficult.
>
> (MK#45)

An Idea of the Time

Earlier, we explored the 'Great Man' ethno-theory, which suggests that one great man invented genre and several entrepreneurs subsequently diffused it. Another common narrative argues that inventors are a product of their time, and that social structures and material constellations form part of the reason why innovations come into being. This theory is exemplified by TED, which was founded by Richard Saul Wurman in 1984. TED was created in order to bring people who work in, or research, technology, entertainment, and design together, to give important and life-altering talks.

> TED conferences bring together the world's most fascinating thinkers and doers, who are challenged to give the talk of their lives (in 18 minutes or less).[23]

The first TED Talks took place in California, and technology was their subject matter (technological devices like the Sony Compact Disc and the Apple Macintosh were shown in some of the first few talks). As such, TED Talks have a strong association with Silicon Valley, the famous scientific area near San Francisco. The initial aim of the event, which was to create a space for the exchange of ideas between experts, evolved into a broader mission in 2006, when the entrepreneur Chris Anderson became the curator. Anderson's mission was to bring science together, and to change the world through the communication of ideas. Subsequently, the slogan of TED Talks became 'Ideas Worth Spreading'. Amongst many other initiatives, Anderson established TED Talks video podcasts, so talks became available to watch online, for free. One other major change that occurred involved the application process: initially, speakers at TED Talks were special guests who had been invited to speak; today, however, anyone can apply to present a TED Talk.

> We believe passionately in the power of ideas to change attitudes, lives and, ultimately, the world. So, we are building a clearinghouse of free knowledge from the world's most inspired thinkers, and a community of curious souls to engage with ideas and each other.[24]

In 2012, the organisers of TED announced that TED Talks had collectively been viewed more than one billion times online.[25] In its 30-year history, many famous personalities like Bill Clinton and Bill Gates have given talks, and TED has also been established in many different countries around the world. TED Talks have become a huge forum for brilliant thinkers, so it is not an exaggeration to say that TED Talks have been one of the most important developments in science communication in the 20th and 21st centuries.

TED is not the only important development in science communication to have emerged in recent years. A movement similar to TED called Café Philosophic was founded in Paris, France in 1992 by the philosopher Marc Sautet. Sautet (1995) wanted to engage audiences in a public event in which philosophic and scientific ideas were discussed. This proved popular, and the movement grew. One of the most important offspring of Café Philosophic was Café Scientifique, which was set up by the television producer Duncan Dallas in Leeds in 1998.

> Café Scientifique is a place where, for the price of a cup of coffee or a glass of wine, anyone can come to explore the latest ideas in science and technology. Meetings have taken place in cafes, bars, restaurants and even theatres, but always outside a traditional academic context.[26]
>
> So public engagement with science is bound to increase in many forms over the next decade.[27]

The establishment of Café Scientifique in the UK was part of a wider movement that grew in relation to PUS.[28] For this reason, Duncan Dallas commented that Café Scientifique was an 'idea of the time'. When Café Scientifique was set up in 1998, Dallas did not know that it would spread across the world. Yet spread it did, largely due to Dallas's article on it that was published in the journal *Nature*. One person who read the article was John Cohen, a Professor at the University of Colorado, who went on to establish a Science Café in Denver, USA in 2003. By 2016, there were Science Cafés in over 150 cities worldwide. Café Scientifique inspired many people, and a number of other similar events were established, all of which aimed to transform scientific self-presentation into new contexts, and to enable communication between non-scientific people and experts. Such events included Science Showoff, Bright Club, and FameLab in the UK; Science in the Pub, Smart Night, Secret Science Club, and Entertaining Science Cabaret in the USA; and Philosophy in the Pub and Fresh Science in Australia, among many more. The Science Slam should be seen as the German equivalent of the many other science communication events which have been established around the world since the 1990s.

Players in the Field: Justifications and Legitimacy

From a theoretical perspective, institutional structure has to be grounded in normative and cognitive legitimacies (Berger and Luckmann 1967). In the previous section, I questioned whether the Science Slam can be viewed as a form of communicative innovation. I will now investigate, in more detail, the experiences of slammers and participants at Science Slams. More specifically, I will explore the various expectations, norms, and values held by slammers, all of which support my thesis that the Science Slam is a novel genre of science communication. I have already explained how rules about timings, scientific content, translation, as well as the need to create an atmosphere and invoke an emotional reaction from the audience form the basic frame of the slam. The next step is to look at why those involved take part, and this information will primarily come from quotes from interviews with participants. This will illustrate the inner perspective, which should provide an ethnographic frame for my video analysis. After, I will characterise the typical inner structure of the genre.

Knowledge Communication

Organisers have identified two issues with the current communication processes in slams. The first problem is that there is a communication training deficit, especially at universities. The training that students receive on communication is subpar and infrequent, which is a structural problem of education. The second problem is that most scientists' communication skills are lacking, which is more of a personal competency issue. Communication education varies across different universities and institutions, so although communication skills are generally lacking, some institutions understand that communication is important and, therefore, provide communication training and have 'fair scientists' (MS#48). Yet, the majority of universities still 'see rather less value in communication' (MS#48). When I asked some organisers what academia could learn about communication from the Science Slam, they had a few comments.

> Pfft, to be brief, I think that it is very important, as we see again and again. Um, well, at scientific meetings there are not . . . many non-scientific people in the audience. You might not be able to explain . . . a topic easily, but I think it is very good if you can express yourself well. It is good if you can sell yourself.
>
> (MK#43)

> Peep out of the box. Stop hiding behind technical terms, do not just talk so beautifully smartly. Look at things with a twinkle. Don't always see yourself as the centre of the world.
>
> (MS#44)

> Go out on stage people, you deserve it. It's cool what you're doing.
>
> (MS#44)

> The academic world ought to learn that you should be able to explain what you are doing to everyone.
>
> (MK#46)

> I believe it is a bit about intelligibility . . . scientists can learn from the Science Slam that you can communicate science understandably. And the second point, which is also important, is that science is allowed to be entertaining.
>
> (MS#47)

> I believe that if a slammer goes back to university and gives a talk, they will definitely perform it differently. They have learned something. The science slammer, in a very short time, learns how to reduce complexity, and how to bring something to a quick conclusion. And then, of course, they will have leant how to ensure that the public does not fall asleep. . . . Therefore,

I believe that there is something positive in being on stage at the Science Slam as a scientist. . . . At least I've never heard the opposite from a science slammer.

(MS#47)

Science Slam organisers wish to motivate young researchers and provide them with training, so they are comfortable addressing and communicating with an audience. Organisers claim that the need to effectively communicate with non-scientists is increasing, particularly at technical universities. Technical universities often have links to external contacts in industry, so being skilled at communication is sometimes necessary in order to secure a job after graduating.[29] It has been suggested that students from the humanities are less eager, or less willing, to learn communication skills because they are generally less dependent on external funding for their research. In fact, many who study the humanities receive money from the Deutsche Forschungsgemeinschaft (German Research Foundation—commonly referred to as DFG) so 'contact . . . with the outside world is a little less, and therefore the pressure is lower' (MS#48). Many organisers agree that funding is 'not driving the humanities to the stage' (MK#45) because students from the humanities have not yet joined 'the game for funding . . . so, they do not feel that it is their job to go public' (MK#45). The fact that the Science Slam is dominated by scientists and engineers may be because they have a pre-existing awareness that they need to communicate in order to receive funding. In interviews, organisers repeatedly contend that slamming is an entrepreneurial activity.

So, there are some people who want to represent themselves on stage or, in any way, sell their concept. Of course, there are also many who have researched . . . developing products. But that is legitimate, in my opinion. That's just a research approach like any other. Also, sometimes in the end they say: 'This is the product that will come out of it'. Indeed, it is a kind of advertising. That is also completely OK.

(MS#47)

Many organisers believe that scientists have a duty to appeal to, and perform for, those who provide funds in order to raise money for their research. One newspaper article about slams addressed this need.

Stefanie Mayer had fun at the gig, and now she knows that she still needs to work on her stage presence. Because next time it might not only be about the show, but about . . . funds . . . which is a daily struggle . . . for young scientists.[30]

The founder of the Science Slam, Alexander Deppert, initially struggled to find people who were willing to speak at slams and, therefore, he allowed slammers in the beginning to talk about anything, even about research they had not done themselves. Later, organisers became much stricter and insisted that slammers could only present their own original research or knowledge.

> In the beginning, I did everything alone. For two years, I was provided only with limited . . . time. Back then I did not insist so much on slammers using their own research. Maybe there was someone who passed through and was stupid, but I could hardly avoid it. I made my life very difficult, because some have come to us since then and said, 'My buddy here has found out something great. Can I present that?' and I have to reply, 'Nope, you cannot, but your buddy can present it.' They then say something like, 'Yes, but the dude has no desire' and I must reply, 'Yes, then, I am sorry.' Yes, so I have lost some willing participants because I tried to be strict. But I think that is one of the benefits the slammer has now, that everyone in the audience knows that they are presenting their original research. Whether that was made clear, or not, is another question. If the audience asks in retrospect, 'Was this really his?' then perhaps this has not been made very clear. But, in principle, it should be known that every slammer has presented something original, and that is cool.
>
> (MS#44)

On Science Slam websites, the most popular term used to describe the Science Slam is 'scientific knowledge'. Other popular terms include 'field of knowledge', 'research', 'scientific topic', 'scientific findings', 'scientific content', 'science', 'wisdom' and 'expert', all of which are frequently used to describe the Science Slam. Slammers can present seminar work, a bachelor or a master's thesis, or PhD research in slams, as long as the knowledge is original. As one might expect, authorship and ownership of knowledge is very important in slams. Nevertheless, this does not stop the Science Slam community from worrying about whether there is enough original knowledge in slams. In fact, if a slammer does not have enough original research in their slam, they can be excluded from Science Slam Championships.

> Martin Buchholz is the first winner of the Science Slam Championship in Germany who presented something that he had just presented in his lectures, and that's fundamental knowledge about thermodynamics. But it was still great. Science Slams should not get too close to a degenerated Wikipedia-Show, that would be a little bit difficult. So, the rule is: present your own research! And most people do it. Anything else would be a bit of a shame.
>
> (MS#48)

Although slammers are encouraged to present their own original research, which is often very complex, many Science Slam websites demand that slammers explain their research in simpler terms in their slams. Terms like 'understandable presentation', 'easy to understand', 'generally simplified', 'in extracts', 'so that non-scientific people can understand', 'guaranteed to be understandable', 'easy to interpret', and 'so all people can understand', are frequently used on websites.

Correspondingly, organisers also aim to reduce their use of complex and technical language when being interviewed.

> So, I think that's one of the most difficult things, to shrink a dissertation of one or two, or three hundred pages to ten minutes and then to cut so much out that I, as a non-specialist, would understand it.
>
> (MS#47)

Removing complex language is also important to audiences, who report that they would like slams to be 'scientific but staged as normal and funny' (RP#52). In other words, they want something educational, but not too complex. In a small survey I carried out, many audience members stated that they wanted to 'take something home' (RP#51) and 'get an insight into unknown topics' (RP#54).

Placing Knowledge

On slam websites, scientists are asked to forget, or leave behind, their dark laboratories, libraries, and lecture halls in order to create a unique experience on stage. In an organiser's meeting that I attended, knowledge locality was acknowledged to be an issue. It was viewed as problematic because, in the past, some slammers had participated in multiple slams in several different towns, which had given them an unfair advantage over others in a competitive scenario. Organisers agreed that they should pass some new rules in order to stop this behaviour. As such, they ruled that slammers could not present at more than one slam, specifically when competing in a regional competition, and they also banned slammers from participating in slams outside their usual region. However, there was disagreement over whether it was better to have an event comprised of local representatives, or of the best slammers in Germany, and whether slammers who have had regional success before should be allowed to compete again.[31]

Later, one organiser told me that her aim was to support local scientists, because there were, understandably, negative reactions when skilled slammers travelled to regional slams to beat local scientists. Although she was happy that any slammers were achieving success, she did not like professional slammers coming in to beat local participants. I got the impression that the Science Slam had proved to be a positive and supportive force for communicative infrastructure in smaller towns.

Legitimacy in the Public

In the interviews I conducted, it was clear that scientific communication prior to the Science Slam was problematic, and that many wanted to do something about it.

> In Germany, it is rare to have exchange between scientists and non-scientists. A lot of scientists stay in their 'ivory towers'.
>
> (MK#46)

Despite the resounding success of Science Slams in Germany, some organisers and scientists have received negative feedback when talking about their research outside Science Slam circles (some have been called evil, while others have said that people stopped listening to them after a few minutes). Feedback like this makes many scientists feel that they have to permanently justify their research to the public. It has been suggested that if the scientific world continues to spend large amounts of public money on research, or if it is inclined to increase its political influence, it will be necessary for it to increase its communication with the general public.

> I always had the impression that people do not deal with science because science does not search for a connection with them. . . . And I think that science has a duty to communicate, because . . . most . . . research is very expensive and needs plenty of money. . . . A lot of tax money is spent on research and this is why I think scientists must inform the public of what they do with the money . . . society is given all the possibilities that science offers, and yes I think people are obliged to build their opinion and deal with science.
>
> (MK#46)

> I feel that . . . on the one hand, it is right in principle that science just informs, simply because we . . . have to play a greater role in society. So many scientists . . . complain that they are asked very little with regard to political decisions and so on, for example with regard to climate change, renewable energy, etc. They always have an opinion on it, but they are not really part of the discourse. There are individual figures who constantly sit in talk shows, but not the real experts. That's one reason. On the other hand, scientists have to justify themselves, they get immense amounts of taxpayers' money for their research. . ., and I think the citizens have the right to know what is done with it. . . . That is why science communication is fundamentally important.
>
> (MK#45)

There is a certain expectation that scientific communication with the public should be high quality because scientists often seem to come from an unknown world. On the other hand, it has also been suggested that members of society have a duty to educate themselves on scientific knowledge. Science Slam organisers predominately believe that it is the scientist's duty to communicate effectively with the public.

> We are there, we are relevant, and we can no longer pull out and say that 'We have nothing to do with you, we are doing our research just for us'. Therefore, we also have a duty to explain what is coming and why we recommend something to the politicians and why there are uncertainties, etc.
>
> (MK#45)

Organisers have two reasons for believing that scientists have more of a duty to communicate their research. Firstly, scientists are the experts and, secondly, as the public funds scientific research by paying tax there is an element of responsibility.

> Basically, most research institutions are financed by the state. Accordingly, I find that people . . . have a right to know what is happening.
>
> (MS#47)

> All have the right to learn what science does and ensure that science gets communicated to the public.
>
> (MS#47)

Sharing scientific knowledge with the public allows scientists a certain amount of reflexivity, but scientists must be able to consider how their work affects, and relates to, the public. In the 17th century, the collective public act of watching scientific experiments was deemed necessary for scientific research to be legitimised. Today, most people already believe and trust in science and scientists, in both an institutional and personal sense. So, scientific research is shared with the public nowadays, not because the public disbelieves scientific research and wants proof, but because the scientific community is financially dependent on the public. Thus, when scientists are introduced as researchers at slams, audience members expect that their claims and research will be scientifically legitimate. It was no surprise, therefore, that when I interviewed a man from the audience at a slam and asked him about his expectations, he replied with the following comment.

> I expect that I will hear about an interesting topic, that is not just jabbering, but there is something funny about it. I think it will also make sense so I will take some knowledge home.
>
> (RP#50)

Even though slam audiences are mostly made up of young, educated people, the task of science in a communicative sense is to achieve a kind of outreach. Indeed, organisers constantly think about recipients, locations, and outreach.

> Lectures in the Haus der Wissenschaften are absolutely legitimate. You just have to be aware that you only reach a specific audience there. You will not motivate the students, who are mostly collecting credit points, and you cannot persuade school classes to go to these lectures either.
>
> (MK#45)

Converting Science into Entertainment

In addition to advising scientists that they should present original research only and that their slams should be as simple as possible, Science Slam websites also indicate that slams should be entertaining. Phrases and words such as 'entertainment', 'show business', 'stage', 'spotlight', 'entertaining', and 'interesting' are commonly used on websites to describe this aspect of the Science Slam. On one website, the Science Slam was described as a 'smarter' alternative to a television show, while another stated that presenting accurate scientific knowledge was not as important as providing an entertaining and informative slam. A third website stated that slams essentially turn 'science' into 'entertainment' and this was echoed by an organiser, who claimed that the Science Slam is essentially a mix of science and entertainment, calling it 'sciencetainment' (MS#45). In a survey taken before one event, audience members were asked what they expected from the slam (the majority of them had never been to a slam before). Most of them said that they were attending the slam because they wanted to be entertained by scientific research. As one audience member put it, slams should be 'exciting, entertaining, but nonetheless informative' (RP#51). The topic of entertainment was addressed by one organiser, as follows.

> If we look at all the people who are entertaining enough to be on stage, not all of them are researchers. And the pool is bigger. Not all scientists are entertaining, but a few are. But there are also people who appear on stage who are not really entertaining but are cool scientists. Or have something to say. And . . . so I keep hearing, 'It must be entertaining'. It is nice when it is entertaining, but there is no rule that states we have to tell people who are not entertaining to leave. I remember a lecture given by Thorsten Fischer; he is our space man. And his lecture was a big one on his stay in space. He showed the audience some really cool videos, but he had a shitty physicist's humour, which I thought was embarrassing. And there was someone in the audience who was obviously a physicist as well. He told a joke and Thorsten Fischer laughed, but it was just not funny. But the lecture was simply thrilling, it was cool. But it was not so much entertaining, because entertainment is usually equated with humour. Let us just say, it does not always have to be humorous or funny. But it should keep people in a good mood and thrill them, so they do not switch off after one, two, [or] three minutes and say: 'What kind of loser is this person?'
>
> (MS#44)

Experience (as a Product)

Just as there is an emphasis on providing an entertaining performance, slammers are also asked to ensure their slam is emotional as well. Science Slam websites use words and phrases like 'exciting', 'thrilling', 'haunting', 'enthusiastic', 'fun', 'joke', 'lust', and 'science can be happy' to encourage slammers to create a special, evocative atmosphere in their slams.

> So, you should, in some way, build an emotional bond or create a relationship! Either through laughter, humour, tension or . . . affection, do something to hook the audience.
>
> (MS#44)

There are a couple of benefits of making a slam emotive. Firstly, the content of the slam is more exciting for the audience. Secondly, by making a slam more emotive, scientists have the opportunity to show off their charisma and 'transform the discipline into rock'n' roll' (MS#412). Organisers believe that the public is interested in both the scientific content of slams, and in seeing the more empathetic and feeling side of scientists. Therefore, organisers ask slammers to humanise science by sharing their feelings, thoughts, and perceptions with the audience. In other words, the slammer should show the audience the 'real person' behind the scientific research.

> And if you make . . . people enthusiastically listen . . . it is absurd—the topic is science and 500 people who would normally have nothing to do with it, are listening in a room. If someone would have told me this . . . ten years ago, I would have said: 'I can hardly imagine that'. Today it has almost become commonplace in Germany that slams take place, people go, and these people are interested in science.
>
> (MS#47)

Finally, the reward for participating at a slam can be experience or monetary based.

> In the IdeenExpo there was a Science Slam. I do not know if it still takes place. But there was suddenly a prize of €5,000 for the winner. That was shitty. Because in all other Science Slams there is basically nothing to win, winners get a book voucher or similar. Suddenly, there was this huge sum to win which was a little weird. The other slam organisers did not like it because they thought it could cause a rivalry, or at least make it difficult for them. Organisers thought, 'Why would slammers perform at our event when they can win €5,000 at another Science Slam?' That was a fear. But slammers usually participate . . . because they enjoy it. This IdeenExpo slam was only open to natural scientists and that was also an exclusion.
>
> (MK#43)

Competition

With regard to the competitive aspect of slams, various terms and phrases are used on websites such as 'winner', 'beat the competitor', 'the audience is the jury', 'success', 'compete against each other', 'champion', 'sporty contest', 'judging', 'fight for the audience's favour', 'the best', and 'battle of the brains'. Traditionally, at the end of slams, the audience has the opportunity to react to presentations by applauding, whistling, hooting, leg stomping, or by using assessment cards. Assessment cards, in particular, can determine who wins the slam. Some audience members only judge slams based on their scientific content, rather than judging both content and entertainment value, because they believe that, in a similar way to competitive diving, if it is a difficult and complicated jump (or topic), then that should be acknowledged.

> One should keep that in mind. Yes, so if someone has a simple theme, it is perhaps not so difficult to make it comprehensible. And if someone has a really complex topic and has to use a lot of analogies and an object in order to make it comprehensible, then this is, of course, of higher value.
>
> (MS#48)

Ultimately, the audience votes for the winner. Most organisers believe that this healthy competition is a central part of the event, as demonstrated here.

> Competition motivates people. It motivates participants to give a good talk because they know that they will be assessed on it. And it corresponds with the desire of the audience to judge, to participate, to interact. It is this searching-for-the-stars thing. That works well, too, because humans are like that! This is why the audience are keen to come.
>
> (MS#48)

Organisers argue that competition is necessary because, according to them, people enjoy judging others. In fact, some even claim that people listen to academic talks much more when they get to judge them afterwards! The process of judging slams also allows audience members to collaborate in order to come to a decision. For example, one organiser commented that he had seen people at slams say something like 'hey, it is nonsense what the group over there has evaluated, the slammer deserves more than that' (MS#48). As with all things, people have differing opinions and so evaluating slams gives audiences the opportunity to interact with people who have contrasting views. In a way, the Science Slam can be compared to sports like soccer, where fans argue over whether play was foul or not, often for weeks after the game! The act of judging slams or adding a competitive element also 'increases the tension' (MK#43). However, voting is not always impartial or fair, because some slammers bring all their friends to their slams to vote for them. Despite this unfair tactic, it should be noted that all the competitors sit and have a drink together afterwards, so it is 'really more a funny thing' (MK#43), than a proper competition.

> No one actually really doggedly wants to win. Secretly, they probably all want to win [laughs] and so are glad if they do not come last.
>
> (MK#43)

Although many slammers do seem to be motivated by competition, others use their platform to spread important messages.

> This is the exciting thing about it. People often do not participate to win, but to present their topic and communicate an important message. For example,

the young woman that I mentioned before, with the bacteria and the coral reefs, she wanted people to realize that bacteria are not just stupid annoying animals, but that they belong in our world. To always evaluate them negatively would be narrow minded and . . . short-sighted.

(MK#45)

Learning

It has been argued that the Science Slam has a certain educational duty. The goal of 'lifelong learning' ('You Never Stop Learning') is heavily promoted on websites, and phrases like 'teaching', 'instructing', 'sharing knowledge', 'instructional presentations', 'help the unknowing audience to understand', and 'make people smarter' are commonly used to describe the focus on learning in slams. Many people attend slams because they want to 'take small portions of knowledge home' (MS#48), otherwise they 'could just go to comedy shows, but this is a little more than comedy' (MS#48). Despite this emphasis on learning, organisers claim that they are happy even if 90% of the audience views slams purely as entertainment, rather than having a particular focus on learning. In this scenario, any knowledge taken away would be purely accidental.

Entertain yourselves. In the end, you have learned that there are scientific institutions, that slammers research some unlikely topics and deal with scientific questions, and that most of the time the research is useful for something, and the slam was somehow funny.

(MS#48)

On the other hand, some organisers argue that slams do not have a duty to educate because they are not directed or funded by the state or academic institutions. Even if one particular organiser believes that slams should be educational, that organiser should remember that everyone has free choice. It is important that organisers feel like organisers, rather than education providers. Organisers also debate over how educational talks will be received. Some believe that audiences will be so fascinated by a slam that they will investigate the topic further afterwards, while others disagree. Meanwhile, slammers hope that a bit of their talk will stick. 'If there is one core message of my talk that sticks in their heads, then I have achieved my goal' (MK#45).

I think you can only enjoy the Science Slam if you have good general knowledge. . . . And a comprehensive base of knowledge can only be developed through . . . education. It cannot be formed in a ten-minute presentation. That does not work. You must read by yourself and work on a topic. No Science Slam is sufficient alone as a method or action.

(MK#45)

Power to the Audience

Organisers claim that audiences become empowered through attending slams. In particular, the act of judging a slammer's performance is viewed as empowering. Usually, no dialogue takes place between the audience and the presenter in a slam. The audience is primarily seen as the receiver of information, although audience members do sometimes interrupt speakers. The audience are aware that presenters are more knowledgeable about the subject being presented than they are, so communication in slams is, for the most part, a monological, linear, one-way process in which the slammer has the monopoly to talk. However, when listeners do not enjoy a talk, they have the power to show their disapproval, either by giving the slammer a low score in the judgement process, or by not applauding them at the end of their slam.

Platform for Novelty

There are frequent references to 'openness' and 'novelty' on Science Slam websites, but the term 'innovation' is rarely used. However, similar phrases like 'actual research result', 'new innovative ideas', 'creative heads', 'no limit to creativity', 'genius ideas', 'sensational discoveries', 'astonishing results', and 'originality' are commonly used. The takeaway seems to be that 'anything goes'. That is to say, any new ideas or concepts are usually deemed acceptable in slams. For example, the sentence 'everything that fits into ten minutes is allowed' was found on one website. This applies to both devices used in presentations, and to the form in which slammers present their research. Websites inform slammers that they can use any type of software or device they can think of, for example, PowerPoint, a handout, a flipchart, or even handicraft work, or a fake gun, as one organiser suggested! Slammers' performance styles are also open to interpretation, and relatively unrestricted. Slammers are free to sing, dance, play, perform live experiments, or just use a PowerPoint Presentation. The only stipulation is that the form of their slam, and the communication of knowledge in their slam, should be different from traditional lectures.

> The only expectation is that slams are not allowed to be like normal lectures— boring. The slammers have to take some risks. This is what makes these events exciting. Nobody knows how far a scientist will go to make a difference.[32]
>
> The concept is quite simple and is similar to performances by young, literary people. One enthusiastic scientist, sometimes with an affinity for profiling, stands on stage and tries to entertain an audience.[33]
>
> I mean, every one of us can give a presentation at a university, but we both know that it is usually very boring and not what we expect from a science slammer.
>
> (MS#44)

> Even if a scientist is brilliant, if he talks to you in a monotonous singsong voice and is the most boring dude on this planet, nobody will listen to what he says.
>
> (MS#44)

While many traditional science communication events are described as non-personal, boring, uniform, and serious, Science Slams should be casual, short, emotional, enthusiastic, artistic, novel, entertaining, and authentic. The old, ascetic scientific personality should also be abandoned. This rejection of traditional scientific communication is one of the major motivations behind the Science Slam. The Science Slam offers a critique of traditional scientific communication, while simultaneously emphasising the importance of high-quality communication. Interviews with audience members confirm that this is the case, as most people attend slams because they expect to learn something new, whilst seeing slammers use unusual devices or tackling unusual subjects. As one audience member put it, 'I expect that it will be amusing and creative, and maybe I will learn something new that I did not know before'. Creativity is of key importance. In fact, Alexander Deppert called science slammers 'artists' and claimed that slammers have developed artistic personalities because of the Science Slam. Despite this glowing assessment, the artistry of science slammers is thought to be less developed than that of poetry slammers.

> There are similarities in the basic format [of the Poetry Slam and the Science Slam], but the content is colossally different. You can send a poetry slammer on stage three times in a row and they would say something different each time. But you cannot send a science slammer on stage twice because they sometimes cannot even explain two different ideas, because they have just one issue or idea. And it is also primarily not about the fun or the language, but about the content. But if you . . . make it so the content can be understood by the public and so a spectator can understand highly complex issues, and if, at the same time . . . the audience is a bit amused; I think then you have done a lot.
>
> (MS#77)

Yet, many slams lack creativity. This is often attributed to slammers using YouTube to self-coach which can harm their creativity. Another reason for a lack of creativity relates to slammers performing in multiple slams. If the same slam is performed more than once, this may impact how a slammer delivers the performance. In contrast, allowing slammers to participate in several slams with a different topic each time, could allow creativity to flourish. Organisers have debated whether slammers should be allowed to participate in two separate slams with the same topic in a German championship, but after it was pointed out that poetry slammers are not allowed to perform the same Poetry Slam twice in a championship, the Science Slam followed suit.

Interdisciplinary Communication

The Science Slam provides a platform for communication between scientists and non-scientists, but it also allows for interdisciplinary communication between scientists from different disciplines. Many slammers report having had fruitful exchanges with scientists from other disciplines after presenting their slams. Despite this camaraderie, there is undoubtedly a problem with representation in Science Slams—the natural sciences are usually overrepresented, and other subjects underrepresented—a fact which organisers acknowledge. Science Slam audiences expect to hear a broad variety of talks from different disciplines, and so inequality is problematic. Organisers would like all disciplines to be represented in slams. However, as the following quote shows, it is more difficult for those from the humanities to prove their legitimacy.

> If someone comes on stage and tells you that he has a genetically changed mouse and that he infects this mouse with cancer and takes a tissue sample, then nobody in the audience will ask if this is real science. But if someone tells you that he analyses women in English novels from the 19th century, then . . . people say, 'I can read books as well'. For this reason, the humanities think they have to legitimise what they are doing. They have a fear of becoming trivial. So, people from the humanities are less self-ironic . . . a hardcore physicist can say a thousand times that he hasn't had sex—I think scientists can make jokes about themselves in a way that those from humanities cannot do.
>
> (MK#46)

Consequently, interdisciplinary communication is only partly successful. Thus far, the impact of the humanities on the Science Slam has been minimal. Organisers emphasise that all knowledge communication is important, and that the Science Slam should uphold four aims: to place knowledge, to make science entertaining, to make interdisciplinary communication possible, and to foster public learning. We can conclude that it is important for knowledge, of all kinds, to be shared with academics, professionals, and the public, even if some fields are underrepresented.

Changes in the Genre

We will now turn to explore some of the changes that have occurred in the genre since the Science Slam began. The first slammers were students who were invited to recycle their university lectures as slams. Alexander Deppert then officially launched the event and made it clear that slams were to differ from traditional conferences or academic lectures. Deppert introduced several unexpected initiatives, like supplying beer at slams, enforcing an informal dress code, and using a dispersed seating arrangement in order to create a different atmosphere (Hill 2017b). From then on, organisers tried to avoid any, and all, formalities; they chose not to use universities as venues for slams and asked slammers not to wear suits on stage. Furthermore, in the beginning, experts of any subject were invited to take part in slams; it was not just academics. Initially, speakers tended to be of mixed age, and older participants were specifically invited to participate. Nowadays, male scientists of a certain age dominate the field.[34]

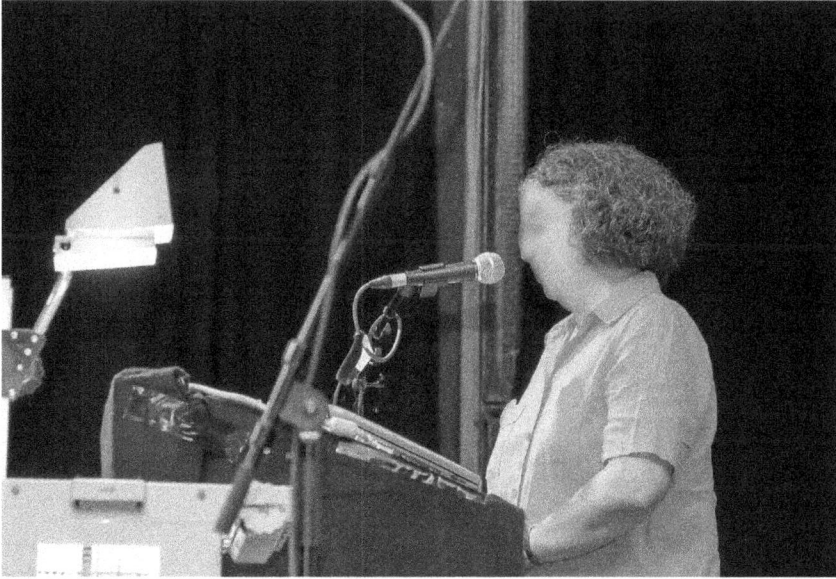

Image 5.3 Anonymous Image From the First Science Slam in 2006

Further differences between slams now, and slams then, concern presentation format. Back in 2004, slammers tended to use overhead projectors and spoke from a lectern. Sometimes, they read from a piece of paper. Over the years, visuals improved in quality, became more significant, and were used more frequently. Presenters, too, have gone from being considered to be text-spouting 'talking heads', into full bodies in communicative action, who perform and orchestrate multimodal presentations. Additionally, the use of media such as YouTube and the practice of live broadcasting slams have further complimented the face-to-face interaction that occurs in the Science Slam. In a way, audio-visual media re-sensualised the presentation of scientific knowledge in slams.

The use of media in slams can be traced back to 2008 where the Haus der Wissenschaften used visual tools to advertise and promote their first slams. Their advertisements comprised a brain logo, which then became the brand logo of the Science Slam. In addition, the use of presentation software, for example, Power-Point, Presenter, and Prezi became common around this time, and this allowed slammers to bring visual elements into their slams. Today, organisers ask participants to 'consider bringing funny little pictures or other punchlines' (MS#49) to their slams. Thus, when preparing their presentations, slammers frequently use Google Images to find pictures that will illustrate their research.[35] All of this has resulted in organisers being very aware of the impact that a good presentation with images can have. In interviews, organisers often enthuse about slammers who have visually impressive presentations because they believe that the use of pictures distinguishes the Science Slam from older, perhaps less technical,

Image 5.4 Anonymous Image From a More Recent Science Slam

Table 5.1 Changes in the Science Slam Since It Began

Year:	2006	2016
Subject:	General Knowledge	Academic Knowledge
Presentation style:	Recycled university lectures	Expectation that the form has to be new and different
Presenter:	A 'Talking Head'	Full bodies, with communicative action
Material:	Paper-oriented	Slide-oriented
Venue:	Not at universities	Any possible venue
Clothing:	No suits	No dress code
Researcher:	Experts of any knowledge	Producers of academic knowledge with an artistic spin
Length:	15 Minutes	10 Minutes
Subject of Research:	General Knowledge	Natural Sciences/Academia
Presenter's age:	Young and old participants	Young scientists
Outreach:	Situated performance	Situated performance & YouTube
Location:	Hall	Hall and Public Viewing

communication forms. Colourful pictures have the potential to rouse feelings and grab the audience's attention, so slammers who use pictures in their slams are considered to have a competitive edge. In contrast, the lack of success and visibility in the Science Slam by humanities students can possibly be explained by their refusal or inability to work with visuals (Hill 2017a).

These changes suggest that the Science Slam has evolved into a new and updated genre, where the demand for different forms of communication and content are met. Good content is scientific, based on original research, and presented in a creative or artistic manner. Bad content is uncreative, based on non-original knowledge, which has been copied, or is presented like a university lecture.

Conclusion

Adorno and Horkheimer argued that the cultural industry gives the impression of being innovative but does not offer any new ideas or theories. The exclusion of anything 'new' or 'novel', at the time of Adorno and Horkheimer, was seen as an indication that a culture was 'infecting everything with sameness'. Habermas later criticised the general public and their lack of communicativeness, describing them as staged, one-sided, noncritical, and undemocratic. I have avoided making similar pessimistic statements about contemporary culture, because I have tried to focus on the ways in which those involved in science communication have talked about novelty and told their stories, instead. This chapter explored how the Science Slam was set up by Alexander Deppert, primarily because he viewed previous forms of science communication as problematic. Deppert believed that the socio-material arrangement of traditional academic lectures, and the way in which the scientific persona was staged in such lectures, needed changing.

Next, we explored the 'Great Man' ethno-theory, which purports the idea that one great man invented genre and several entrepreneurs disseminated it. We also explored a theory which suggests that inventors are a product of their time. A look at recent science communication events outside Germany suggests that the Science Slam, and other similar events that have emerged since the 1980s, are not just the story of a single great man but are the result of particular socio-cultural conditions that have helped shape them. I believe that there is no single factor that explains innovation. Rather, innovation is the result of a confluence of factors that must be accounted for when discussing it. Innovation in the Science Slam can be viewed, I believe, as being the combination of both an event and of particular material and technical circumstances of the time.

Despite its general success, the Science Slam has not escaped criticism. Resistance to the Science Slam predominately comes from established scientists, who think that the Science Slam is not the correct way to communicate science, and that sharing scientific research with the public is not necessary. Some poetry slammers have also reacted negatively to the Science Slam due to their belief that the Science Slam 'stole the format' of the Poetry Slam. In response to these criticisms, organisers emphasise how much effort and risk it has taken to make the Science Slam a success. Organisers acknowledge that other events and inventions have had an impact on the Science Slam but argue that no one should have copyright over communicative arrangements.[36] So when critics condemn the Science Slam's practice of borrowing elements from other genres, organisers reply by highlighting the novelties of the slam.

In the introduction to this chapter, I argued that, in experience-based societies (Schulze 1992), the joy of having adventures is based on the alternation between embodiments and the reception of embodiments. If the experience economy focuses on consumers who want subjective experiences and surprise, then the call for a permanent novelty of form (Hutter 2015a) has consequences for communicative genres. The reflexivity of the Science Slam must, then, be influenced by novelty of form and aesthetic experiences. Therefore, concerning novelty, we can conclude that on the one hand, novelty appears in the form of new research and, on the other hand, novelty is the freedom to choose any type of device to assist in making a presentation.

This motif of novelty appears consistently in many areas of the Science Slam. From championing inventors and entrepreneurs, to website descriptions and the rhetoric used in slams, participants and organisers alike believe that their event is a new, novel form of science communication. The reasons that slammers give for wanting to participate in slams also show novelty. Some want to solve societal problems, others want to share their scientific research with the public, but most of them simply want to challenge the traditional 'typical scientist' image. Novelty is present, too, in the expectation that slammers should communicate differently. Rather than following the communication methods used in traditional lecture formats, slammers are encouraged to communicate differently in their slams. So, although the Science Slam can be viewed as derivative of academic lectures or the public experiment, it must be acknowledged that it has developed its own way of communicating. This is most likely the result of social structures and materials, as well as technological developments and cultural leanings, all of which have made the Science Slam possible.

From a theoretical perspective, it makes sense to describe knowledge as a social reality, a guiding principle for action and, finally, as a tool with which to relate to other members of society. For this reason, genre knowledge should always be considered in its historical context. By grounding my analysis in a historical framework, it has become apparent that 'new' genres are caught up in a tense relationship between institutionalised ways of communicating science (reproducing social structure), intending to do something new (the call for novelty), and communicating differently (creative action). In the case of the Science Slam, I have shown how both references to established structures, and the motif of novelty as a relevant orientation for action, support the idea that the genre represents an innovation of scientific communication.

Notes

1 Godin (2014a, 2015) stated that from the Reformation to the 19th century, innovation was a political concept that was viewed negatively and forbidden by law. 'The valorisation of innovation probably started after the French Revolution, but the real paradigm shift to innovation discourses across society happened in the late 1940s to early 1950s' (Godin 2008, 2014b, 21). Daston and Park (2006) argued that if the people in the '"Age of the New" [The Reformation] had been "[. . .] asked to give their own epoch a name, they would perhaps have called it "the new age" (aetas nova). New

worlds, East and West, had been discovered, new devices as the printing press had been invented, new faith propagated, new stars observed in the heavens with new instruments, new forms of government established and old ones overthrown, [. . .]' (Daston and Park 2006, 1).

2 On the one hand, Schumpeter showed how an increase in productivity through innovation was possible, but on the other hand, he also exemplified how innovation could destroy established industrial companies when a user accepts new, technical goods (or new methods of production, new markets, new supply sources, or a new organisation of an industry) instead of simply replacing older ones. This, in turn, leads to an ongoing demolition and re-creation of economic structures. For Schumpeter, the process of creating new goods was driven by new combinations because producing something new involves combining materials and forces in a different way. Although the new combination may, in time, replace the old one through continuous adjustments, there is certainly change and possible growth, but neither a new phenomenon nor development in our sense. Development in our sense is defined, then, by carrying out new combinations (Schumpeter 2011, 51).

3 Gilfillan (1935 [1970]) applied many of Ogburn's theories to processes of technological innovation. Gilfillan believed that innovation was the result of new combinations of established knowledge.

4 In addition to Ogburn's overreliance on evolution, another issue with Ogburn's work is that he described society as non-material and he separated technical developments from social developments.

5 I think that social innovation researchers would do better to speak of 'humanitarian' or 'socially caring' innovations. From a sociological perspective, there is no non-social innovation and, therefore, this should also not be a characteristic of any innovation. I also recommend viewing social practices as characterized by materiality, and I stress that it should not be the primary motivation of the researcher to detect whether innovation actually leads to improvement, or not. I think it would be good if sociological research on innovation did not pursue a 'pro-innovation bias', but rather always acknowledged the social dimension in innovation and placed emphasis on its relational nature.

6 Similarly, Hutter et al. (2011) proposed a theory on innovation that dropped the pragmatics, semantics, and grammar of innovation and instead sought to develop a new reflexive model of action in social spheres. This new concept of innovation suggested that innovation is continually discussed in public discourse (semantics), but is not usually innovative in practice (pragmatics).

7 Knowledge can be built and shown in different forms of objectivation. Linguistic and objective signs are comparable forms of materiality and should therefore be equivalent base for innovation research.

8 'The central idea of communicative constructivism is that everything that is relevant about social action has to be communicated/objectivated (observable and experienced): by objectivation, we refer not only to objects produced but also to the body's performative activities and the corresponding objects addressed, involved, or produced, such as sounds, gestures, or facial expressions; they also include objects referred to, objects produced, signs, and technologies. Particularly with respect to innovation, it must be stressed that, as part of communicative actions, objectivations always imply a certain meaning' (Knoblauch 2014, 6).

9 Haraway also argued that fiction is important. 'Fictions can be imagined as derivate, fabricated versions of the world and experience, as a kind of perverse double for the facts or as an escape through fantasy into a better world than which actually happened' (Haraway 1989, 3).

10 This is comparable to Joas' (1992) idea of 'situated creativity', wherein actions are not only drafts, but also answers to situations.

11 Luckmann (1986) distinguished between the internal and the external structures. The internal structure, according to Luckmann, was related to patterns of genre and explained its basic functions (in which problems of communication were solved). In this, material manifestations like prosody, rhetoric, language register, topoi, phrases, gestures, facial expression, and rules of interaction were important. The external genre structure was concerned with the interrelatedness of communication and the institutional setting.

12 Unlike Günthner and Knoblauch (2007), Yates and Orlikowski used evolutionary rhetoric and structuration theory to describe genre development. I will describe the dialectic relationship between social production and the objectivated world slightly differently, based on social constructivism.

13 This argument is based on the idea of the interpretive paradigm, which states that structural conditions of action exist, but that actors are not victims of these structures (Hildenbrand 2009, 33). Structures do not determine the plot of action, rather, actors can choose between alternatives.

14 Are these re-framings similar to Goffman's (1974) description in his book *Frame Analysis*? Goffman's thoughts on re-framing can be demonstrated by his example of a door-to-door salesman who transformed cleaning with a vacuum cleaner. The salesman's cleaning of the floor does not really have the effect of cleaning the floor, rather it shows somebody how cleaning a floor would work.

15 All interview extracts were translated from German to English.

16 'My boss responded positively to it and gave me time, so to speak, to implement my idea' (interview with Alexander Deppert, 2014).

17 The original quote in German: ‚Oft halten Studenten hervorragende Referate, leider aber nur ein einziges mal. Bei uns können sie wiederaufbereitet werden' (Article "Packende Vortäge, gereimte Zweizeiler," Darmstädter Echo, published December 12, 2007).

18 Another organiser put it like this: 'In the beginning, I just said the talk has to be comprehensible and ten minutes long'.

19 The original title in German was ‚ *'Packende Vorträge, gereimte Zweizeiler'*, Darmstädter Echo, published December 3, 2007).

20 'So, when you look at the super successful slams, well, YouTube is a bit different. On YouTube it is a certain rule of law that even if one gets many clicks on their online video, they may not be successful in front of an audience' (MS#49).

21 Since many audiences often show up in large numbers, several organisers set up a public viewing platform where people can watch a slam take place while sitting in a separate room. So, slams transcend both through the medium of YouTube and also through live broadcasting.

22 The slammers in the final were Anastasia August from Karlsruhe, Simon McGowan from Hannover, Kai Jäger from Bonn, Helga Hoffmann-Sieber from Hamburg, Franca Parianen from Leipzig, Victor Lopez Lopez from Bochum, Lei Mao from Berlin, and Simon Reif from Munich. The slam was moderated by Andreas L. Maier and the winner of the 2013 championship, Reinhard Remfort, was the featured scientist.

23 Source: accessed January 14, 2012, www.ted.com/pages/about.

24 Source: accessed September 19, 2015, www.ted.com/about/our-organization.

25 Accessed January 15, 2013, http://blog.ted.com/2012/11/13/ted-reaches-its-billionth-video-view/.

26 Accessed January 15, 2014, www.cafescientifique.org/index.php?option=com_content&view=article&id=72&Itemid=484.

27 Accessed January 17, 2014, www.nature.com/nature/journal/v399/n6732/full/399120a0.html.

28 When looking at the recent history of science events in Europe, it is important to bear PUS in mind, because it was connected to many events.

29 'Well, every scientist should do something. It deserves to be standard training, at least for doctoral candidates. Yes, every man who is a doctorate candidate should have at least one week of communications training. That would be a great goal' (MS#48).

30 Accessed July 20, 2015, www.rbb-online.de/stilbruch/archiv/20150219_2215/science-slam-berlin.html.

31 In another sense of the word, moderators frequently discuss the local area in their dialogue at the beginning of slams. They often give facts about the town the slam is taking place in, as well as the town where the slammers come from. Local knowledge and spatial awareness are important topics, for both slammers and organisers.

32 Accessed August 20, 2015, www.zeit.de/wissen/2010-02/Science-Slam-2010.

33 Accessed August 20, 2015, www.spiegel.de/wissenschaft/mensch/science-slam-wie-eine-liebesnacht-den-raum-kruemmt-a-675595.html.

34 Initially, Deppert was happy for any speaker to present a slam, and he invited his friends, poetry slammers with academic backgrounds and extroverted people he knew to take part. Presenting original knowledge was not crucial, nor was using only academic knowledge.

35 Several slammers have become infamous for using refined comic-style illustrations in their slams. In 2011, graduate student and cartoonist Kai Kuhne illustrated his slam on '*Political trends in German labour law*' by turning it into a cartoon. In 2012, Giulia Enders' presentation '*Gut with Charm*' incorporated illustrations created by her sister, Jill, a communications designer.

36 TED talks are often described as in delineation.

6 Science Slam as a Genre

Since its conception, the Science Slam has been widely regarded as a new genre of non-academic science communication and described as 'communicative innovation' (Byers and Mead 1968), because it both establishes new knowledge and represents a new form of communication. The Science Slam is often seen as an alternative way of legitimising and communicating science. To better understand processes of communication and to show how they are socially embedded, the focus of this chapter is on analysing successful Science Slam presentations (or those which are labelled 'new' or 'better'). This analysis must be seen as an acknowledgement of the theoretical focus on embodied subjects, objectivations, and the relation to others, which has previously been explored in Chapter 3.

Broadly speaking, genres of non-academic science communication include three aspects: an **internal structure,** a **situated realisation,** and an **external structure**. The internal structure is viewed as an overall blend of different elements like words, phrases, registers, rhetorical figures, styles, and rules of interaction. In my analysis, I will try to identify certain language characteristics, such as guiding motifs and topoi, while also commenting on the media. It is worth mentioning that in addition to these characteristics, a linguistic aspect is often expected. In Science Slams, this comes in the form of a performance by a speaker in front of an audience and includes gestures, facial expressions, and interaction tools. Next, I will look at the way science is embodied in public science communication. Science Slam performances are both partly situated and partly situating, which means that some parts of a Science Slam are socio-technically prearranged and other parts are constructed in communicative processes. Finally, my analysis looks at the external structure which encompasses the relationship between communicative action and social structure. Social structure includes themes like financial dependency, extra-institutional settings, social milieus or norms, and rules of communication.

Preliminary Science Slam Encounters

In this section, I will share some musings from one of the first Science Slams I attended. This occurred soon after I was offered the chance to start a PhD on new approaches of public science communication. I heard that there was a Science Slam taking place in Lido, not far from where I was living in Kreuzberg Berlin. It

DOI: 10.4324/9781003172635-6

Image 6.1 Giulia Enders at her Science Slam

was 27 February 2012 and I decided to attend the slam with my mother Rosemarie Sanyang-Hill, who had come to visit me.

As soon as we entered the location, we noticed that there were several cameras in the room. The event was packed with people and we only found one free seat in the first few rows, which my mother was kind enough to give to me. The event was hosted by the slammer André Lampe, and participants included Garcia Peters from the University of Hamburg, Falko Brinkmann from the University of Münster, Peter Westerhoff from the Charité, Berlin, and Giulia Enders from the Goethe-University, Frankfurt.[1] The slam began with a presentation by Garcia Peters, which was followed by the second talk, presented by Giulia Enders. Lampe introduced Enders to the audience and gave an overview of her academic qualifications, before requesting that the audience welcome her with applause. Enders then began her talk.

GIULIA: *okay, as mentioned before, I am studying* medicine.
>   Voice from the audience: *Woohoo*
>   (Cheers)
GIULIA: *[laughs]* (.) yes, that's right. and one thing *that medicine is really good for, like you can see right now, is drinking coffee.*
>   (Looks to the slide. The slide shows three aunts holding teacups)
>   *or tea with aunts.* ah, because *when drinking tea you are often asked what you are studying and while my sister has to explain for half an hour what communication design is, I just say medicine.*
>   (Looks to the slide. New slide shows smiling aunts.)
AUDIENCE: [Laughs]

It was clear that Enders felt very comfortable talking about her discipline, which she, and clearly her aunts, saw as a worthwhile field of research. The following remarks, which she made after visiting her aunts, made it clear, however, that her specific research subject—the human gut—was not quite as popular as the broader term medicine. After she talked about tea with her aunts, she explained that her interest in the human gut was the result of a conversation with her roommate in which he came into the kitchen of their flat one morning and asked her to explain pooping. In her talk, she elaborated on the subject of the gut and spoke about topics such as the bowels, farting, sphincter muscles, and bacteria. Unusually, she used several informal words like 'shit' and 'pooping', which caused giggling and laughter in the audience. Enders explained her arguments using visual jokes and comic-style illustrations that her older sister, Jill Enders, had drawn for her. If anyone in the audience assumed that scientists were boring, lab-coated white men, Enders, and her talk, would have made them question this assumption. Enders' authentic, casual, and unembarrassed style was unconventional, but her charming and easy-going nature allowed her to easily handle the, perhaps unpopular, topic of the gut. A newspaper article published after the event described the video recording of Enders' talk as such:

> In her YouTube video, the girly, pretty student Giulia Enders talked about one of the last taboos and great mysteries of the human body—and she even went into the hard Biomolecular details.[2]

When I attended the event, I did not know how iconic Giulia Enders would become or how big of an impact her presentation would have. She won the slam that night for her presentation '*Darm mit Charme*' (Gut with Charm)[3] and shortly after, a video of her slam was uploaded to YouTube and became an instant hit. Although she had only participated in four Science Slams, her videos went on to have more than one million views (1,088,800 as of July 2014). Enders also received an offer from Ullstein Verlag (one of the largest publishing companies in Germany) who asked her to write a book about her research. Shortly after the book was published in March 2014 it sold more than one million copies and took first place on the Spiegel bestseller list in the non-fiction paperback category. As a result, Enders was invited onto talk shows on television and she became prominent in the newspapers.

In my ethnographic experience since, many Science Slam participants acknowledge that they have seen Enders' slam. Some even admit that her slam motivated them into participating in a Science Slam themselves because they were inspired by her style of presenting, which had an impact on their own presentation. Enders' slam could, therefore, be viewed as a catalyst for many future slammers, and slams. On a personal note, Enders' talk served as inspiration for my own fieldwork and research.

I do not intend to primarily focus on incredibly successful science slammers like Giulia Enders, but instead on science slammers who are successful in several situated Science Slam settings. However, it is important to start with Giulia Enders because she had such a big impact on the public image of science, and her slam marked my own entry into the world of Science Slams. Of course, she is exceptional but to focus on her is to focus only on how one slam can hypothetically establish new communicative practices in science. She is a great example of how traditional ideas about science can be challenged and of how scientists can be successful in both a professional setting and in 'popular' culture, two areas which have often been described as 'conflicting'. Overall, she embodies the hope that public science communication can break the traditional conventions in science communication, for example, norms, gender, technical jargon, and visual practices.

The External Structure

Venue and Setup

On Science Slam websites, scientists are asked to leave behind their dark laboratories, libraries, lecture halls, or desks in order to create a real-life experience on stage. Spatial metaphors are often used to establish a boundary between science and the Science Slam. Despite this, certain expectations about space and time are expected and are the norm. These include the expectations that events will take place during the week and start at either 19:30 or 20:00 and that the entire event will be two to three hours long with a break halfway through. In addition to this, strict rules dictate that the individual presentations should be no longer than ten minutes. Science Slam venues, meanwhile, are heterogeneous and slams can take place in traditional venues like science centres and theatres, or in less traditional venues like rock concert stages and jazz clubs. More recently, the use of universities as venues has become common.[4] One thing the venues seem to share is that they are not specifically related to science. The following quote from one of my interviews exemplifies the variety of venues that can be used.

> It begins with churches, goes to cinemas, theatres, jazz clubs, bars, clubs, i.e., everything that in the broadest sense can serve as a venue. And of course, it also has to work with the operators, which has to do with the conditions, with which the place you can get. . . . You have to get a relatively cheap place because otherwise you cannot perform the event. Or the minus is much bigger than it would have been anyway. So therefore, you're trying to find places that have a certain cultural, scientific affinity, or that make clear that there is no goal here to make money. People, who make Science Slams, basically are people who are highly motivated to carry out an event. It hardly ever comes to economic factors. That is why one tries not to maximise profits, but to minimise cost. If I can save money on venue, then I'll do that.
>
> (MS#47)

Even in the slams I attended, there was huge variety in the venues selected. One particularly unusual and historic location was Bogen 2, a vault which sits in the bridge pillars of Cologne's Hohenzollern Bridge through which you can see the Rhine and the Zoobrücke through the windows. The vault space itself comprises one large hall, with space for about 500 people. The tattered space combined with the dark lighting gives Bogen 2 an unusual charm. Bogen 2 is usually used as a club and as it sits below a train station, you can feel and hear when the trains above move over the tracks. Next door to Bogen 2 is a charity for homeless people.

To get an idea of the atmosphere present at many event locations, I will give an example from my ethnographic experience at the German Championships at the Festsaal Kreuzberg.

> Almost all the seats are occupied. There are 260 seats available, but if there were no chairs in the room 600 people could easily fit. Today, just a few people stand. The room is divided into a lower and upper level. On the other side of the stage there is a bar, one of a few indicators that the room is usually used for concerts and parties. Other signs include the disco ball in the middle of the ceiling, and the stickers and graffiti all over the walls. I hear loud music coming from nearby such as Thomas D, Bob Marley, Nelly Furtado, and Die Toten Hosen. . . . Lots of people are talking and discussing in the audience, and in every row roughly a few persons hold a beer, including the slammers who sit in the reserved front few rows. On the stage, preparations are still ongoing, a man stands preparing something on a computer, cameras are being put into position, and the lights are being set up.
>
> (MH#22)

Most Science Slams are full of people. Loud, youthful music is often played and drinking alcohol is an important aspect of the event. In fact, some organisers have told me that they sometimes make deals with the owners of the venues so the price for hiring the venue is cheap because the money earned from the bar sales makes up for it. The stage is usually full of media and there are lots of cameras around.

Some Science Slam locations are like those used for Poetry Slams and concerts; others are more like lecture halls in universities. At most events, however, certain key elements are immutable. The science slammers almost always stand on a raised stage, while the audience usually sits on bier benches or in theatre-style chairs on the ground and some people stand behind the chairs. Technical equipment usually consists of a laptop, a projector, PowerPoint software, a microphone (either in the slammer's hand or on their head), a laser pointer, a remote control for the projector, spotlights, and video cameras. Most of the time you have to pay an entrance fee to gain access to the event. Science Slams were initially dominated by a few individuals who organised events, but the genre has now become more established in an institutional context. As we have already explored, large number of slams today take place in universities, company buildings, schools, television studios, and at trade fairs.[5]

Finances

Science Slam organisers use various means to generate revenue for their events. Some organisers portray themselves as modern entrepreneurs:

> I had never seen a Science Slam until I organised one. I had only read this report. I had never seen one. . . . This was also reflected in the newspaper article of Heinz Fischer. He was a guest at my first Science Slam. He wrote something like 'His approach of unimagined boldness' or something or 'thoughtless". . . . My parents were shocked. How could this boy have organised something with no experience? And a friend who works on a doctoral

thesis on innovation and self-employment and the differences of young and old people called me immediately and said: 'Yes, super. These are the entrepreneurs of today. Just do it. That is what is important'.

(MS#44)

Other organisers have a tendency to mention their own impact on the event.

So, the only thing we did—we just turned it into a competitive event. I'd say Alex Dreppec developed it at the beginning of 2006 and I think at the beginning he was not so sure about the competition; not as explicit as it is today.

(MS#47)

Science Slam organisers could, of course, be described as entrepreneurs. However, this does not mean that they make enough money to work solely as organisers. In fact, it is hard for most to make a living from slams and so many organisers have second jobs.[6] These jobs include leading science seminars and giving lectures or workshops on science communication. Only a small minority state that organising Science Slams is their main job. In some cases it is possible to have a part-time position funded by the Ministry of Education but most claim that external financial support is necessary for one to hold a slam in Germany.

This financial support can come from multiple different sources. One common way is for organisers to coordinate Science Slams in the context of conferences or be employed by the state or museums, who provide financial rewards for holding slams. Another is to use a sponsor. Sponsorship can come from various sources: the local city or university, banks, insurance companies, and magazines. It is important to note, however, that organisers often state that they would not work with sponsors who would manipulate an event in their favour. In fact, slams organised by companies are often demarcated from 'real' Science Slams, for example, those organised by companies such as Haribo, IdeenExpo, and Shell. Although the influence of sponsors does generally seem to be minimal, it cannot be denied that they have some effect on the slams they sponsor.[7]

Now I cannot think of many organisers in Germany of Science Slams who make it just for the glory or for, how do you call it, for the good cause. Really, I do not have many in mind—so either they are instructed to carry out something, because they, I'd say they have a student or a PhD position and are paid anyway. Or they do it for a theatre or run this event for an external event venue. And then you have just a few that do it differently. And they need to search for supporters. Otherwise, it is not possible. Because based on the entrance fees you cannot realise a Science Slam. That will not do.

(MS#47)

In the slams I attended, various financial sources were used. In these slams, the entrance fee varied from nothing to €7. Some slams had sponsors and these included companies like Geo Magazine, the job website 'academics.de', the University of Hamburg, Radio Q, the bank Sparkasse Münsterland Ost, Münster Marketing,

BASF, Junge Köpfe, and Weitblick e.V. Other slams were funded by the German Federal Ministry for Education and Research. These slams were mostly free to enter and were always connected to the 'Year of Science' scheme (presentations in this had to be related to topics like Health Sciences or Digital Society).

Milieu

Contextually, Science Slams do not have a place in classical bourgeois culture, but rather fit into the context of a certain communicative milieu. The participants of this communicative interaction can be described as a discourse community, a speech community, or a communicative milieu (Swales 1990; Hymes 1974).

Broadly speaking, organisers can be divided into two groups; the science dropouts who have gone into business for themselves and science communicators who work for an organisation. One quality they all seem to possess (at least, all the ones I met) is an academic background. Their backgrounds are diverse and include subjects like German Philology, Cultural Studies, Theatre Studies, Politics, Public Policy, Management, International Management, Architecture, Physics, Biology, and Technical Environmental Support. Many of the organisers who were science school dropouts told me that they began their careers working as research assistants in scientific institutions or at universities. Some even worked for parliament before they decided to establish their own business and move into freelance work, which they labelled as 'risky' and 'daring'. The other group typically started working straight after university in a science communication job for a specific company. Interestingly, many of the organisers had counselling or public relations experience prior to organising Science Slams. With regard to gender and age, organisers tend to be fairly split between the sexes and are usually aged between 30 and 40.

These statistics do not ring true for the Science Slam milieu as a whole, though. According to organisers, those who attend slams are usually young, sophisticated, and committed. Correspondingly, the Facebook statistics from one slam indicated that most of the visitors to the page were between 25–35 years old. Similarly, by reducing the entrance fees for students, another organiser was able to calculate that one third of the audience at her slam were students. Aside from students, most organisers thought that the rest of their audience were educated people interested in science. Other attendees include colleagues and pupils of the research groups run by the organisers, and people from creative industries (for example, graphic designers, web designers, programmers, and journalists) who live in the 'hip' neighbourhoods where many of the events take place were expected guests as well.[8] For this reason, one organiser labelled the Science Slam audience as a 'gentrification audience' (MK#45). In other words, highly educated, young, and upcoming.[9]

Situated Realisation

The Role of the Host

A typical feature of slams is the role of the host or moderator, who introduces the slammers and then presents the rest of the event. In the following segment, I will analyse Science Slam introductions and give an overview of the communicative setup on the background of Conversation Analysis (CA) (Sacks, Schegloff, and Jefferson 1974) and ethnographic knowledge. The slam I will be analysing is from the Science Slam Federal Championship, set in Berlin. The question I will try to answer is how the layout of an event is set in a Science Slam. To begin with, I will give my ethnographic impressions of the socio-technical arrangement. Then, I will describe the opening sequence, before turning to discuss how the structure of the event showed situated action.

The Science Slam Federal Championship, organised by the Haus der Wissenschaft, Braunschweig, and funded by the Federal Ministry for Education and Research, was created for the 'Year of Science' in 2014. As part of this, scientists were asked to present research from the field of 'Digital Society'. In the lead up to the final (July to November 2014), slammers competed against each other in preliminary competitions held in Braunschweig, Bonn, and Saarbrücken. Those who came in first and second place qualified for the final in Berlin. At the final, the top six slammers were asked to compete with a brief presentation. The event started at 20:00 at the Fritz Club in the Postbahnhof in Berlin. The event was called the 'Federal Final' (Bundesfinale), not the German Final, because it was funded by the Ministry of Education and Research.[10] I will now give a description of my experience at the event.

At the entrance to the Postbahnhof, there is a red light where The Fritz Club is. I wonder if it used to be a station building. I hope there is space at the event because in the last few years the event has been full. I can already see the queue, but it only contains around 50 people. A young man says, relieved, 'The queue is okay'. A few minutes later he adds, pleased, 'It's moving!' The entrance is brightly lit. One of the people waiting says anxiously: 'Oh, they take the drinks bottles' while another one says, 'The drinks are confiscated'. At the entrance, a young woman has to give her packet of milk to security. Inside the venue there are postcards on the walls. On the right, there is a coat check which costs one euro per item. The walls are white painted brick and cables visibly protrude from the ceiling. In the great hall, many people sit on their jackets on the ground. I see fixed cameras in three places around the room. There is also a mobile television camera controlled by a cameraman. On stage sits a very large screen displaying a slide which reads 'Science Slam in Science Year of the Digital Society'. I hear ambient music. A guy plays with his iPad and says 'Get me a beer! How much does a beer cost?' Another guy replies: '€4'. The guy with the iPad answers, 'Then I don't want one, that's too expensive'. Then, with a laugh, the other guy adds 'Fetching it was included in the price'. Both laugh. Another young man tells a group of four smiling men something about the moderator Stefan Raab. One young man crudely massages another guy who groans loudly in pain and laughs. The girlfriend of the guy being massaged comes back from the toilet and asks incredulously 'Is he giving you a massage?' She then sits down next to them. Shortly after, she tells her boyfriend and the masseur 'That's enough'. There is a smell of alcohol in the air, a fragrance of beer. Some people roll cigarettes; others look at their smart phones, while others chat whilst having a beer. The room gets increasingly packed. Two women who look around 50 and who have short hair and wear suits stand out from the crowd. A boy around 13 who is accompanied by a lady who looks around 45, also stand out. Most visitors wear casual clothing (hooded sweatshirts, jeans, colourful t-shirts, woollen sweaters, hats, and scarves). The audience seems to be composed of equal parts women and men, while the average age is probably around 25.

(MH#22)

The description above details the preparation and opening sequence of the event. I will now explain what happened next. The moderator of the event was the actor Andreas L. Maier. When the event started, he walked up to the heightened stage with a microphone in his hands.

The slam begins with the moderator walking on stage while the audience applauds. Before beginning his dialogue, the moderator looks briefly at the ground and then at the audience. He remarks that he is happy that his fans are there (in the audience). This comment could either be an ironic remark, a joke, or his usual style of welcome at slams. He makes his first announcement and asks those who are 'sitting in front so comfortably' to 'stand up for us all' to make room for those outside the event to come in. By doing this, Maier seems to pressure the audience into being compassionate. He thanks those who have stood up and asks the audience to applaud 'everybody who is STANDING UP'. The audience applauds and Maier then comments on the sudden change of mood. He uses a nasally voice as if imitating a member of the audience and says, 'We expected seats'. Some people reply and say 'Exactly'. The moderator returns to his normal voice and says, 'We didn't have an entrance fee . . . listen, with these prices the moderator had to carry chairs by himself'.

As we can see from this, a socio-technical arrangement came into focus. I can further elaborate on this situation using my ethnographic knowledge. What is not shown in my transcript is an interaction that occurred prior to the moderator's introduction. In this interaction, one of the organisers asked those who were sitting down to stand up so more people could be let into the venue. This was met with irritability and one guy in the audience even shouted, 'I find it rude that no chairs are offered!'

Despite the man's protest, around half the audience reacted to the organiser's request and stood up. In the front few rows, however, only a few people moved while some people who had stood up decided to sit down again. All the while the people sitting discussed why they thought they should not have to stand up. At the same time, new people were arriving and pushing to the front row. A guy in the first row said, 'Here there is no space, only one space'. Many people voiced their complaints about the lack of seating, one guy said, 'Standing two hours is really hard'. A small number (15 people) stayed sitting. This whole scene was recorded by the cameraman, who focused on the audience with his video camera. The reason a lot of people reacted so badly to having to stand up was because Science

Slams usually have seating, and it was unusual for people to be asked to stand. People who arrive early to slams usually get a seat.

After his opening remarks, the moderator began his regular introduction. The moderator addressed the audience in a deep, nasally voice. He stated what day it was and then asked the audience how they felt. After a cheering reaction from the audience, he said that he wanted to hold a small test and asked, 'Who actually is a Berliner?' After counting some raised arms in the audience, he concluded that 'There are exactly 18 Berliners'. In my experience this is a typical introduction to a slam. It included reference to the town or location, and questions about how familiar the audience is with the event. This lead to a clarification of the slam, which is also usual. The moderator described Science Slams as similar to Poetry Slams, but without poetry. He paused briefly for three seconds to allow the audience to react. Then he jokingly said that he understood this is hard to comprehend but that the audience 'must understand this'. The moderator proceeded to explain slams in more detail. 'It's a genre that was actually invented in Germany in 2008.'[11] He remarked 'Hey' afterwards in a protracted, high voice which seemed to be a way of appealing to the audience and resulted in a cheer from them to which he happily replied, 'Very good, you are all present'. Next, the moderator stated that slams are usually only performed by 'more or less young people' who have done 'their own research'. He put more emphasis on this phrase by saying it in a loud, protracted way. Then he stated that people who think a topic is cool, but who have not done research on it are not allowed to present at a slam. He assured the audience that the six slammers presenting tonight had done their own research. Yet again, he shouted 'Hey' in a high, protracted voice. Again, this drew cheers from the crowd. After a small interlude in which he explained that he is from Cologne where people are much more enthusiastic, he calls the Berlin audience 'tough sledding'.

The moderator explained that each presentation should be ten minutes long and that after the presentation it was the audience's turn, 'then it is your turn', to discuss what they thought of the slam, and decide if they liked it or not. The phrase 'like it' was spoken with a different, special-effects-style voice and Maier explained that he asked Thilo in the sounds team to alter his voice so that whenever he said 'like it' his voice would sound different. He says 'like it' twice more and the audience laughed. Maier continued and said, 'The man is good' and then says 'like it' for a third time. He continued explaining that the audience must make a decision about the presentations 'This means er. . . . YOU decide, if you liked it or not'. 'You' was emphasised. Maier suggested that the audience's judgement should not be just a simple thumbs up or thumbs down (this could be a reference to social media) but based on thoughtful discussions with one another. Maier then suggested criteria the audience could use to judge the slams. These included whether the audience liked the slam, if they understood the slam, if they could explain the slam to their grandma, and if she would also understand it, and finally, if the slides design and presentation fit the content. After weighing up these criteria, the audience should assign the presentation a point between one and six.

In this chapter, I've explored what one can expect at a Science Slam venue and what the atmosphere is usually like. I've also explained what the role of the host or moderator is and I've looked at one specific slam and how it was run.

The Interactional Organisation of Science Slam Presentations

Historically, in public science events, the audience and the speaker have interacted together through face-to-face communication. In a university lecture, for example, speakers give a long, uninterrupted presentation usually followed by a discussion. Although the Science Slam still involves face-to-face communication, it is different from a typical lecture. In a Science Slam, the speaker gives a short presentation which is followed by an immediate judgement from the audience about the content and form of the talk. In general, after hearing a talk the audience intensely applauds or unenthusiastically stays quiet. The slammers sometimes ask the audience short questions. The traditional lecture form is, however, present in the arrangement of the space and the rules governing who has the right to speak in a slam.[12] What differs between a slam and a lecture is that a slam involves much more spontaneity, simplicity, and less formality than a usual lecture.

Science Slams also include expectations about people and the roles they play. People who go to Science Slams expect to be part of a multi-sensory, one-to-many, face-to-face, linear type of communication format. They expect a mediating presenter (moderator) who will introduce the slammers and subsequently order the event. The audience know that the slammers compete against each other, so they expect the slammers to have representational skills, as well as an ability to demonstrate and share their research by using various modalities like body movement, language, and projected slides. Finally, the audience expects to celebrate scientists or slammers by applauding and cheering.

In the following part, I will present the usual order of a Science Slam.

a   In the *opening sequence*, the moderator introduces the speaker. The speaker/ slammer then comes on stage and prepares to present his or her talk. The slammer introduces themselves and tells the audience about their research.

b   In the *orientation sequence*, the slammer explains why they are interested in their research (e.g., it could be for scientific thinking or solving a societal problem). This part often starts quite generally and ends with a more specific focus. The *orientation sequence* allows the presenter to show and demonstrate to the audience that he or she is a scientist and a '**do-er**' of science.

c   The *solution sequence* involves clarifying an ambiguity in the speaker's research or clarifying the role of the researcher. The solution sequence is designed to create surprise so regular translation processes are part of this.

d   In the *evaluating sequence*, the presenter reflects on their research. Translation processes are a regular part of this too. In this sequence, the overall relevance of the research (either societal or scientific) is given.

Inner Structure

Showcase for Interactional Organisation

I will now explore a successful slam in order to demonstrate the typical sequence of a Science Slam performance. This slam took place on 6 June 2013, at a youth club called Komma in the town of Esslingen. The slam was organised by the Office for Cultural Affairs of Esslingen (run by Stefanie Bayer), and the Hochschule Esslingen—University of Applied Sciences. The slam was moderated by the actor Andreas Laurenz Maier, who is from Cologne.[13] The entrance fee was €5. The presentation in question was by Johannes Schildgen, who has won over 20 Science Slams since 2012. Schildgen is a researcher in the field of informatics at the University of Kaiserslautern, a role which involves him analysing huge amounts of data. Schildgen is infamous for two different Science Slams and we are going to focus on one of these which is called '*Do you want to have fries with your fries?*'

The slam begins. The moderator is dressed in a contemporary style and wears a hat and red shoes. He speaks loudly and informally, and he reads from a card. As is typical for Science Slams, the moderator introduces the speaker by giving a bit of information about their academic background and tells the audience where they are from. In this instance, the moderator informs the audience that Schildgen attended the Technical University of Kaiserslautern where he studied computer science and that he has a doctorate. After this introduction, Schildgen walks on stage and prepares to start talking. He walks over to his laptop and opens his presentation. The first slide of his presentation displays the title of his slam, his name, his academic title, his email address, the logo of marimba (a programming model that he developed), and the corporate logo of his university. He also wears a headset and holds a remote control in one hand, while the other hand is in his pocket. His talk starts with a typical sequence of scripted interaction.

Schildgen begins by saying 'Hello, who has ever been to a Science Slam?' He puts his hand up to protect his eyes from the spotlights and looks into the audience to see who has raised their arms in response. Then he asks the audience who of them has ever ordered something online. He counts down from three and makes them say in unison where they have shopped online. Whilst talking, he leans over the audience as if anticipating their reply and they reply by shouting 'Amazon'. This makes Schildgen laugh as though he had been expecting this response. He answers, 'Yes, I'll talk today about Amazon' and then he starts explaining why Amazon is such an interesting topic.

Orientation Sequence

What follows this is a short orientation sequence in which Schildgen introduces the area of interest—Amazon—and asks, 'Why is Amazon so successful?' According to Schildgen, the answer is because it seems to stock everything you could ever need and it has free shipping. Schildgen muses over other reasons that could explain Amazon's success. He believes it could be partly due to Amazon's suggestion function. He follows this up with a few jokes. Firstly, he focuses on the absurdity of Amazon, specifically, as he says, the fact that delivery is offered before you have even ordered! Secondly, he jokes about the review tool, which he exemplifies with the example of Adobe. He shows a screenshot of a comment by a user on Amazon who has rated Adobe, 'My wife is beautiful again and I am now fearless in showing her picture to others'. The audience laughs. His next joke involves asking the audience what it means if a user rates a product with just two stars. A guy in the audience replies that they would be 'an unsatisfied customer'. Schildgen jokingly answers 'Two stars, that would be a satisfied Swabian' (Swabians are a group who live in Baden Württemberg).

Schildgen gives an example of the humorous suggestion function, by telling the audience that those 'who bought a World of Warcraft game were recommended Clearasil'. Schildgen then shows a slide with a screenshot of a World of Warcraft game on Amazon, with a suggestion on the side of the screen for things 'you might also like', with a picture of a bottle of Clearasil. This joke is both verbal and visual. Schildgen explains that Amazon makes 30% of its profits by offering customers products that they did not initially search or ask for.

This leads to Schildgen's key argument, which is that he wants to explain how the advertising strategy on online platforms like Amazon and Spotify can be used to sell chips at a chip shop. He argues that if Amazon can advertise in this manner, then he, and the audience ('we'), can also do this. Next, he shows a slide with a picture of an old lady who works at a chip shop standing by a chip pan, which makes the audience laugh. He says the audience should try to imagine that they want to recommend products to customers in a chip shop and to think about how they could do this. Schildgen describes how one could do this and begins with a 'very primitive approach' in which he suggests one could offer mayonnaise for customers to try. Whilst explaining this, he shows a slide with a picture of the same old lady with a speech balloon above her saying 'Try mayonnaise'. He comments, 'Great suggestion isn't it?' in an ironic voice.

After highlighting another unsuccessful strategy, Schildgen informs the audience how Amazon successfully suggests products. It does this by analysing the purchasing behaviour of customers in order to identify a certain checklist containing a list of what each person buys. So, if fries are bought, then Amazon would recommend that they try other fries. The old lady appears on the next slide, again with the speech balloon, but this time it reads 'Try fries'. Schildgen argues that this is still an unsuccessful technique because a consumer would have realised this by themselves.

Schildgen further explains how Amazon uses the suggestion technique to encourage consumers to buy additional products. Amazon does this by combining products in their 'checklist' in a two-dimensional space. Schildgen illustrates this by showing a picture of fries, a hotdog, and a hamburger on two axes and explains that Amazon adds strokes to their checklist if a customer buys one of these products. In this way, it is possible to generate a huge list of potential additional products. When applying this to a chip shop, Schildgen explains that when a customer comes to a chip shop and orders 'fries without everything', then the chip shop lady would take the 'product matrix' and multiply it by the 'fries-vector'. He says that the result would be the fries-vector on top because the number on the top, the highest number, is allocated to fries. Immediately after explaining this, Schildgen comments in an unusual voice 'Do you want to have fries in addition?' He finishes his talk by explaining that he named his talk 'Do you want to have fries with your fries?' because, although it may sound stupid, this is how product suggesting is carried out at Amazon.

Solution Sequence

In the solution sequence segment, the central idea or argument of the slam comes to a conclusion. By highlighting the many failures of Amazon's suggestion system, for example, suggesting products one has already bought, Schildgen clarifies his position as a developer of these functions, although somewhat more successful.

Schildgen suggests a modification of this system. He argues that we should develop a system based on different customer types and create statistics related to each type of customer, for example, the businessman, the student, the hippie. He says that if we use this approach, it is possible to create a customer matrix. So, if a 'hippie' comes into the chip shop, the lady in the chip shop can multiply the customer matrix by the 'hippie-vector'. Schildgen explains that the music service Spotify uses a radio function similar to the suggestion system he recommends, in order to help improve Spotify's music suggestions. By using key, speed, and melody, Spotify suggests music similar to the music the customer usually listens to. To demonstrate how this works, Schildgen attempts to play the song 'Somewhere Beyond the Sea' on YouTube. However, YouTube does not allow him to play this song and so he says 'Okay, what do we do now? Should I sing, or what?' As if planned, a microphone and a ukulele are passed to him and he sings several songs which share a similar melody ('Somewhere Beyond the Sea', 'Das Meer wiegt sich im Wind', 'Always Look on the Bright Side of Life'). This demonstrates how Spotify uses its suggestion system.

Evaluating Sequence

In the final sequence of his slam, Schildgen reflects on his talk and speaks about his research in more detail, asking 'But what actually is my research?' He then explains to the audience that the suggestion systems are only one part of it. According to Schildgen, services like Amazon and Spotify 'wish to be analysed and that takes an extremely long time' because 'data is subject to change', which means everything needs to be recalculated. This recalculation can take days because Amazon has millions of different products and customers. Schildgen's approach, in comparison, allows one to directly apply changes to the data. However, this takes longer than a simple suggestion service because it is very complicated. Schildgen takes this opportunity to stop because he is running out of time.

Schildgen's research and suggestion system, although in theory scientific, is arguably not motivated by science. Schildgen provides two primary motivators for his interest in this topic. Firstly, there is the economic factor, or more specifically, profit, and second is the failings of the contemporary system. Ultimately, Schildgen shows that he can improve ways of analysing product data, which helps to improve services like Amazon. To put it another way, his role is to improve the inefficient suggestion system.

Humorous Interactions

Science Slams are often filled with humour. Although organisers emphasise that joy should not be the sole emotion triggered by performances, humour is an important part of many slams. Speakers plan their talks with the emotional reaction of the audience in mind and most slammers leave short breaks for laughter after amusing parts of their slam. In the following few pages, two humorous interactions will be explored. In the first example, the slammer presents himself in a comedic way.

The example comes from a slam on health-related research (*'Science Slam im Wissenschaftsjahr'*) which took place on 16 November 2011, at the Festsaal Kreuzberg. The event was organised by Haus der Wissenschaften and the Ministry of Education.[14] The slam was videotaped by the organisers (I was at the event but took notes instead of videotaping it).[15] The slam in question was presented by André Lampe, a Physicist and PhD student at the Institute of Chemistry and Biochemistry (Membrane Biochemistry and Molecular Cell Biology) at Freie Universität (FU) Berlin.

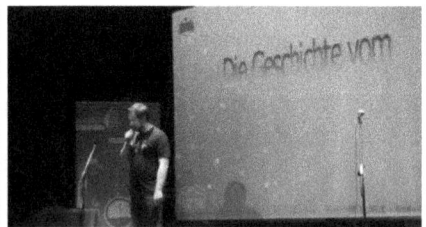

Image 6.2 Andre Lampe's Science Slam   *Image 6.3* Andre Lampe's Science Slam

(Image 6.2)   (Image 6.3)

> G'DAY (2.0) **(image 6.2)** yes. as=uh (.) was already said (.) my NAme is
> André Lampe. (1.0) I am a PHYSICIST (1.0) **(image 6.3)** and I come origi-
> nally from BIELefeld. (1.0)
> (shifts from one leg to another; legs move easily).

Image 6.4 Andre Lampe's Science Slam

Image 6.5 Andre Lampe's Science Slam

(Image 6.4)   (Image 6.5)

life (.) *has not been very kind to me.* (2.0) **(image 6.4)**
(turns his head to the left, looks at the floor and moves two steps to the left)
[Audience laughs]
like that was my research life- um (1.0)
(He turns around and goes back to his starting position. He pulls the microphone cable)
yes **(image 6.5)** *it has had a somewhat slow start and then faded out=uh* (.)
(steps from one leg to the other, raises his right arm briefly)

Image 6.6 Andre Lampe's Science Slam

Image 6.7 Andre Lampe's Science Slam

(Image 6.6)   (Image 6.7)

definitely you always have the problem as er=scientists that you have to JUS-
Tify your research to your funders. (2.0) i had to do that as well (1.0) very
recently. (2.0) *as i sat* (.) *with my parents* **(image 6.6)** *at their COFFEE table*
(3.0)
(scratches his head on the right)
and um (5.0)
[Audience laughs and applauds]
boy, what is it actually that you are doing? **(x)**
(in a different voice)

In this sequence, André Lampe humorously presents himself as a financially dependent guy, from a bad hometown, who studies a boring subject matter. It is quite typical for slammers to depend on their parents for financial aid. Indeed, male scientists often jokingly call themselves 'mamas' boys'. As well as portraying himself as a humorous individual, Lampe also gives off an air of modesty. In his talk, he steps from one leg to the other, scratches his head, and looks to the ground while talking, all of which portrays a 'loser' image which he plays with. At the same time, however, he seems quite confident. The way he alters his voice and posture when imitating his mother, the way he pulls the microphone cable when walking, his cheeky smile and the pauses he takes between sentences all give the impression that being on stage is natural for him. His whole style of presentation is multimodal and well-orchestrated. He speaks with a slight dialect and uses a typical linguistic style, though he speaks quickly.

Orchestration and Timing

Science Slam presentations usually consist of a triadic structure between the presenter, their audience, and the material communicated. As the material communicated partly relies on technology, the presenter-audience interaction is often interjected with presenter-machine interactions. These moments are usually well orchestrated and are made up of acts like a presenter clicking on a laptop or a remote control while taking a sideways glance at the PowerPoint slide or the laptop screen. This interjection or 'beat' is a key part of presentations and it is clear to both the presenter and the audience. Presenters use this 'beat' as a tool for sequencing their talk, for example, when to move between the pre-sequence, general introduction, main topic, excursus, results, and closure. The audience use these turning points for interjections, applause, and comments. In fact, timing these interactions in the midst of the presenter-audience-machine relationship is key when sequencing Science Slams.

Another important characteristic of Science Slams is the significance of timing. Most science slammers have rehearsed their slams over and over again, so they know exactly which slide will appear next. I will now demonstrate the power of timing using the following short fragment, which was recorded at a Science Slam in Berlin, in 2011. The presentation was supported visually by slides. Prior to this fragment, the speaker had explained that his parents had asked him what he researches. To answer, the speaker said that he took two weeks off work and went to his parent's house to explain his research to them. This explanation took the form of a story.

Image 6.8 Andre Lampe's Science Slam

Image 6.9 Andre Lampe's Science Slam

(Image 6.8)   (Image 6.9)

> *I took off for two weeks (.) and er=I returned to the coffee table (.) er=after*
> ***(image 6.8) I***
> (moves his hand in a circular motion)
> *er (.) dealt with the indignation,*
> (turns quickly to the right, then back to the front)
> *ahem (.) I told my parents*
> (looks quickly to the remote control nearby)
> *(.) the story (.) of (.) **(image 6.9)** a testicle snapping fish.*
> (presses a button on the remote and points it to the screen, where the word
> 'testicle-snapping-fish' appears)
> *ahem=this*
> (quickly lifts his right arm)
> *(.) actually describes very well what **(image 6.10)***
> (bashful shrug)
> *I did my diploma thesis in*
> (moves one step backwards, looks to the ground)
> *ahem and it is a little animal fable. **(image 6.11)***
> (straightens up, smiles to the audience)

Image 6.10 Andre Lampe's Science Slam

Image 6.11 Andre Lampe's Science Slam

(Image 6.10)   (Image 6.11)

once upon a time

(squints at the laptop. A picture of a fish appears on the screen)

a fish.

[Audience laughs]

The combination of image, language, and body performance is quite clearly recognisable in this sequence. In the first part of the extract, the speaker turns to the slide and presses the remote he holds with his left hand, at the slide (as in image 6.9). As he says 'testicle snapping fish', the words 'testicle snapping fish' simultaneously appear on the slide. This is a great example of how language, image, and performance can be used effectively. Meanwhile, the importance of timing is demonstrated by the last part of the fragment. In this, the speaker says 'Once upon a time' followed by a two second break in which he casts his eye on the laptop in front of him (as in image 6.10). Shortly after, an image of a fish appears on the screen behind him, while he says 'a fish' (as in image 6.11). The speaker pauses to allow this to sink in and is rewarded with subsequent laughter from the audience. We can conclude, therefore, that the act of leaving a pause for the audience to process their reactions is well orchestrated and pre-planned, while showing and telling is a key part of this.

The Loosening up of Power Structures

A typical characteristic of the communicative genre is interaction. In Science Slams, as in all other situations of co-presence, the presenters constantly interact with their audience. One interaction which commonly occurs in slams is scripted interaction. This usually involves the speaker asking questions, which are then answered by the audience. The following short fragment of scripted interaction taken from a Science Slam in Bochum in 2011 shows how easily speakers can feel obliged to repair a situation in order to protect their reputation.[16] As the slam is an example of the significance of scripted interaction, I will focus on the relationship between the speaker and the audience in the slam.

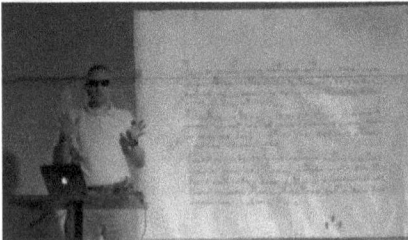

Image 6.12 Schmeh's Science Slam

(Image 6.12)

Image 6.13 Schmeh's Science Slam

(Image 6.13)

SCHMEH: *and (image 6.12) the first thing a cryptologist does, if he analyses an unknown (.) text, is to*
count the letters.does anyone know which letter is the most common in the German language?
(quickly points his right hand to the audience) **(image 6.13)**
AUDIENCE: *E*

Image 6.14 Schmeh's Science Slam

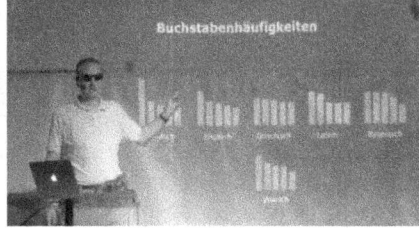

Image 6.15 Schmeh's Science Slam

(Image 6.14)   (Image 6.15)

SCHMEH: *E, exactly, thus cryptologists always say,*
(throws his arms up in the air) **(image 6.14)**
give me an a? (1,0)
AUDIENCE: *E*
SCHMEH: *exactly, and then they say, give me an N,*
(clicks to the next slide. On the black background, frequency graphs appear) *and then give me an I and an S, and if you counted right,*
(points to the silver screen on the side) **(image 6.15)**
then according statistics are the result. And on the left side you can see what statistics in German typically (.) look like.
VOICE FROM THE AUDIENCE: *Step to the side*
AUDIENCE: [laughs]

Image 6.16 Schmeh's Science Slam

Image 6.17 Schmeh's Science Slam

(Image 6.16)   (Image 6.17)

SCHMEH: (Looks to the ground, moves one step to he left. Quickly looks at the
   silver screen on his right-hand side) **(image 6.16).**
   I=hope you can <u>see</u> it. so
   I hope I am more <u>interesting</u> than a <u>statistic</u> **(image 6.17)**
   (points quickly both hands to himself) (laughs)

In the first part of the segment, the speaker asks the audience which letter is the most common in the German language. After the audience collectively shouts 'E' he affirms this, 'E, exactly, thus cryptologists always say give me an a?' Whilst saying this, he raises his arms in the air as if indicating that it is the audience's turn to say something. The audience responds by saying 'E'. Schmeh continues and explains that the most frequent letters after E are 'N', 'I', and 'S'. The slide behind him at this point shows how these letters are distributed. Schmeh points with his left arm to the screen whilst looking at the audience (as in image 6.15) and says, 'and on the <u>left</u> side, you can see what statistics in <u>German</u> *typically* (.) <u>look</u> like'. He pauses for a second and a guy in the audience yells, 'Step to the side'. The audience laughs. The speaker looks down at the floor, takes a step to the left and glances at the screen. He comments 'I I=hope you can see it. So, I hope I am more interesting than a statistic' (as in image 6.17).

Slammers often prepare small question and answer sequences to use as a tool for interacting with the audience. They predominately use simple questions that have short, predictable answers. The fragment I have just analysed is a typical scripted interaction. In this sequence, the take-away impression is that the audience's attention is very focused on the slides. Thus, if the speaker obscures the view of said slides, then the audience speaks up to alert them to this. This indicates that audiences at slams feel comfortable interrupting the speaker. This is a new concept in the genre of science communication, and it is more than likely going to result in many speakers 'losing face', as indeed they have done on some occasions. One reason for this change to the relationship between speaker and audience is that the traditional power structure seen in science communication events in the past has become looser.

This is due to a number of reasons. Firstly, the Science Slam genre is highly interactional and people with different backgrounds demand a share of the researcher's questions and findings. In other words, the power gap, which—according to Goffman—is still typical in the classic lecture structure, has been reduced. This is mirrored in the 'turn-taking' structure, typical in Science Slams. This is in contrast to the classic lecture structure which usually involves a speaker verbally recounting a pre-arranged text and determining when the attendees are allowed to ask questions. In a Science Slam, meanwhile, the presenter's speaking time is decidedly shorter and can, in fact, become a kind of pronounced staccato as the slammer tries to speak without constant interruptions. In this sense, the Science Slam is less one-sided and hierarchical than the classic lecture format. There is, therefore, no room for subject-specific conceit because Science Slams are focused on relationship building and understanding. The difference is that the speaker is the one who asks for understanding or empathy, while in the classic lecture format the listener is the one who is often left struggling to understand.

The 'Rough-Diamond' Style of Science Communication

Science Slam organisers rely on a certain 'in-the-making' aesthetic. It is typical for the audience to see the stage for a slam being prepared while they are waiting for the event to start. Due to this, Science Slams could be described as having a 'rough-diamond' style of communication. Another interesting feature of the genre is the focus on individuality. Coaching offered by organisers is designed so as to not encourage uniformity. In interviews, those who are considered 'good' slammers are those who present their own research and have creative and artistic skills (Hill 2017a). On the other hand, 'bad' slammers are often described as uncreative scientists, who present textbook knowledge that they did not themselves research or those who copy others, thereby encouraging uniformity.

Despite the fact that uniformity is discouraged, and creativity valued, most slammers try to avoid using professional jargon and use everyday vernacular instead, although the use of 'code switching' is common. There is a slight tendency to use slang and Anglicisms. Several slammers address the audience informally using the German words 'du/ihr' ('you'), while others speak more politely. In Science Slams, it is acceptable to have a dialect or accent and many of the successful slammers have slight dialects (most speak with a southern, western, or northern German accent). Some slammers even emphasise their dialect. Although accents vary, most slammers share a certain eloquence in the way they communicate. Most are articulate professionals and only a few have issues expressing themselves. A few of the more successful participants use a style of speech close to fine rhetoric.

Code switching is the act of switching between technical jargon and everyday language and back again on a regular basis. The following segment, which exemplifies this, is from a Science Slam which took place in Duisburg in 2012. As this is an example of code switching, I will only focus on the language used in the talk.

er
(Points to the slide with his finger)
but what interested me especially, were the
frugivore bats, those which nourish on fruit,
(points to the screen with a laser pointer to where a picture of fruit is
  shown)
and here especially (1,0)
(clicks to the next slide, looks quickly at the screen)
artibeus jamaicensis or what we call, *A.J.*
(English pronunciation)
the jamaican fruitbat. (2,0)

> *AJ doesn't do anything <u>all</u> day (.)*
> (English pronunciation)
> *except <u>feed</u> on <u>figs</u>. (.)*
> (clicks to the next slide where photographs of figs are shown)
> *and this—he likes figs so <u>much</u>, that he (.) <u>per night</u> (.)*
> *about of er- receives/eats <u>his</u> body weight in <u>figs</u>. this would be about*
> *ninety <u>kilos</u> of figs for <u>me</u>. (1,0) Er (1,0) <u>not</u> mine.*
> (Keeps both arms apart)

I focus my attention on the beginning-middle part of this extract, where the speaker outlines that he is especially interested in bats that nourish on fruit. After referring to these bats by their scientific name—'Artibeus Jamaicensis'—the speaker says that 'they' (probably meaning himself and his research team) call this type of bat 'AJ'.[17] The phrase 'Bats that nourish on fruit' becomes the technical, scientific formulation 'Artibeus Jamaicensis', which is then turned to the Americanised 'AJ'. This is a prime example of moving from everyday vernacular, to scientific jargon, and back.

Giving 'Body' to Science

It could be argued that science slammers give 'body' to science and knowledge because they are representatives of science, and producers of knowledge. They offer their bodies, their body language, their voices, their clothing, and their gestures. Scientists can be overweight, wear too tight or too loose clothing, their hair can be messy, but they must act authentically.[18] The following summary of Johannes Schildgen's slam indicates how bodily performance is an important aspect of slams.

> Schildgen wears casual clothing, a printed green t-shirt, and black jeans. He has short, dark hair and wears square glasses. His whole presentation style could be described as loose and laid-back. He smiles whimsically throughout and stands on one leg, while jiggling the other. He often gestures feebly using his arms. He seems to be lacking in self-confidence or is shy, perhaps insecure. His accent indicates he is from the Eifel. His voice sometimes sounds nervous and from time to time it cracks. It seems as if he has not planned his dialogue and so he often acts spontaneously. This all gives the impression of authenticity, as though he is being true to himself and not 'putting on an act', which sometimes happens.
>
> (MH#22)

In slams, insecurity is welcome. Some slammers even actively avoid behaving like the 'perfect' scientist. In the following extract, the slammer Boris Lemmer is keen to show that he is a 'real' scientist, rather than an idealised representation of one.[19]

Image 6.18 Boris Lemmer's Science Slam   *Image 6.19* Boris Lemmer's Science Slam

(Image 6.18)   (Image 6.19)

(Holds his right forefinger to his chin)
we will look at a short cut, why? *much is not real, but some*
(keeps his right hand on his chest while his thumb and forefinger form a circle)
parts are.
(glances back at the screen, puts his left arm up in the air)
<u>*THIS*</u> *(image 6.18)* *is a physicist waiting for the first collision of two protons, she is very excited as you can see but she is not real as she wears a* <u>*WHITE COAT.*</u> *If she* <u>*ACTually*</u> *(image 6.19)*
(grabs the fabric of his open check shirt with his left hand)
wore a <u>*CHECK shirt,*</u> *then she would be real,*
(points behind him to the screen)
[Audience laughs and applauds]
(He smiles briefly)

In this extract, the mediatised representation of a scientist is confronted by what, according to Lemmer, a real scientist looks like. Lemmer labels a lab-coated woman a 'fake' scientist, as opposed to himself who wears a checked shirt and is a 'real' scientist. Lemmer touches his shirt while saying that he is the true representation of a scientist.

Visualisation

In Science Slams, the use of software like Microsoft PowerPoint or similar seems to be obligatory. This software is, therefore, likely to become an essential prerequisite of visualisation which has become increasingly important in science communication. Visuals in Science Slams come in a range of different forms including style, pictures, and statistical visual information. I will now examine various types of visual information or content that can be found in Science Slams.

In line with the deliberate avoidance of professional jargon and the emphasis instead on everyday language, contemporary images tend to dominate over scientific ones in most Science Slams. Science slammers frequently use pictures from Google Images—that is, generic visual topoi—to illustrate their arguments or research. These images are often taken from movies (some examples include *Star Wars, Batman, Illuminati, James Bond,* and *The Godfather*); TV (examples include *Pimp My Ride,* a Möbelkraft advert and *Carglass*); from comics (examples include *The Flintstones* and *The Simpsons*); sports (for example from boxing and soccer); food (examples include beer, pastries, kebabs, sausages, spices, beer, and Snickers); of politicians (images of leaders like Schmidt, Bush, Berlusconi, and Merkel have been used); common websites (screenshots of Wikipedia, Adobe, Amazon, Spotify, and webpages of pseudo-scientists were used) and, finally, from the church (images of the Pope have also been used). As well as this comprehensive list, other images include representations of people from different social or ethnic groups (for example, overweight crooners, a black boxer, women in bras, a working-class lady in a chip shop, and old people without teeth). What all these images share, however, is that they are usually used in a comic way.

Translational slides, meanwhile, are often used to bridge the gap between specialised scientific knowledge and common sense. This is done by comparing the scientific images with their metaphorical counterparts from non-scientific life. Scientific images are still a key part of many slams, however. Images of experimental setups, gene codes, magnetic beads, blood samples, nerve cells, glia cells, bones, iron oxide, the brain, and statistical images like diagrams and heat maps are frequently used. Images of the slammer's workplace are common (examples include research areas, a Google Maps image of CERN, a person in hospital, surgeons in an operating theatre, and students in a cafeteria), as are pictures of the researcher alone or the researcher and their team at work. Presenters sometimes like to show images of the equipment they work with (images of scientific apparatus like an MRT or the CERN Hadron Collider, syringes, and results or measurements are common).

In addition to showing scientific images, bringing physical scientific objects to show off and touch is also common. These physical objects become an important part of the triadic structure between presenter and audience. In one such example a presenter pretended that he showed *The Voynich Manuscript,* a book that was written sometime in the 15th century. He told the audience that *The Voynich Manuscript* was handed over to him by Yale University when he asked for it. He held up the manuscript which, from the outside, looks like any other shabby, old book, opened up the book, looked at it, and turned a few pages. He then unfolded a page, held the book up and showed the unfolded side to the audience. The science slammer then touched the manuscript without gloves and gave the audience a chance to look inside the book.

Integrating Science Into Society

Science slammers attempt to integrate science into general society and, in doing so, try to appeal to the average citizen. Slams offer the opportunity for the observation of contemporary communication, and science. I will now share an example which demonstrates how Science Slams attempt to integrate science into society. This Science Slam took place on 20 October 2013 in Bogen 2 in Cologne.[20] This slam was the third round of the German Science Slam Championship and also the final.[21] I attended the event, interviewed some slammers, and videotaped it.

In this slam, the speaker started by telling the audience that he was happy to talk about his beloved research topic, arteriosclerosis. He introduced himself as Johannes Hinrich von Borstel, a PhD student at the Cardiology Clinic Laboratory at the Biomedical Research Centre in Marburg. He described arteriosclerosis and explained that it is a fatty deposit or plaque deposit in vessel walls, which can cause heart attacks and strokes, two of the major causes of death worldwide. Typical sufferers of arteriosclerosis are those who 'love to eat . . . like to smoke, and the typical alcoholic'.

Hinrich von Borstel explained his research further 'So, how do I work now?' He expands and explains that his research involves working with 'C57BL/6 receptor knockout mice'. These mice carry dendritic cells in their bones, which are very important. Borstel's research involves gaining monocytes from bone marrow and loading them with iron oxide in order to prolong the cells as long as possible by finding the best combination of the components. However, he says that it does not matter what combination you use, because it is enough to just throw both things together as the 'boys are tough and just eat the iron'. This made the audience laugh. Borstel illustrated this process by showing three pictures of the same cell but with different amounts of iron, which Borstel described as 'pretty cool pictures . . . almost like in *The Matrix*'.

Next, Borstel explained what he does with the cells. They get put into a syringe, which gets injected into a mouse. After being injected, the mice get chased into a magnetic resonance tomography scanner, where the team hope to get images of them. Borstel looks down. Some people in the audience laugh, while others shout 'Oh'. He admits that he feels sorry for the mice because he thinks they are really cute and 'poor little beasts'. While he said this, the image of a syringe moved towards the image of a mouse on screen. As a joke, Borstel requested his laboratory stop using animals but use brain-dead humans instead (the slide behind Borstel subsequently showed a picture of von Guttenberg, a German politician who lost his doctoral degree after he was caught plagiarising), but unfortunately his colleagues did not like this idea! The audience enthusiastically laughed and clapped again.

Eventually, von Borstel tied these strands of information together and explained that the main focus of his research is to clarify how the aforementioned cells travel, and in this way promote atherosclerosis. Near the end of his slam Borstel reverted back to talking about animal testing. He explained that the one benefit of his procedure versus the 'old' procedure, is that previously, broken vessels had to be extracted from mice, so you would need to kill the mouse if you wanted to look at their vascular system. Whereas, with his method, 'no mice die'. The audience reacted with cheers and claps. He ended by commenting that even if you cannot revolutionise the treatment of people quickly then you can, at least, revolutionise the treatment of mice.

Hinrich von Borstel wore a bright, striped polo-shirt and light jeans. Borstel used slides, interacted with his presentation, and used his body language to expand on his words. He spoke politely, yet confidently, and frequently used both medical and 'non-scientific' language. The audience seemed amused by his descriptions of medical research, for example, 'just throw both things together'. Although Borstel seemed to enjoy his research in general, he clearly does not enjoy testing on animals. His friends, he says, know how hard this is for him, but unfortunately there is no other way to cure human health issues than through animal testing. Fortunately, his new procedure ensures that animals are no longer killed. We can conclude that Borstel has a conscience and he only does the bad (animal testing) for the greater good (human health).

Legitimacy in Science Slams: How Useful Is Science?

Science Slams can be understood as a re-modelling of messy, private, scientific, research. Slammers do not perform scientific research on stage, but instead they create an idealised performance, which participants are aware of. However, all Science Slams, to a certain extent, demonstrate science in use and/or in practice. Slams can be divided into two types—Science and Use and Use of Science.

The first type of Science Slam is what I call 'Science and Use'. In a 'Science and Use' slam, the slammer shows how they would approach a scientific phenomenon and then they argue why this is useful for society. The statement of legitimisation is often paired with a sentence like 'Great, and why do we need all that?' An example of this type of slam is the testicle-snapping fish presentation, in which the unnamed speaker identified that scientists often have to persuade those who provide them with funding of the legitimacy of their research, which is a problem. The slammer concluded by explaining why this research is useful for society. In this case, his research could be used to help people with health problems like Alzheimer's disease. A second example comes from a slam on '*Determining the local heat transfer coefficients at a lamellar tubular heat exchanger by means of infrared thermography*'. After explaining heat transfer and lamellar tubular heat exchangers, the slammer argues that this is important because more efficiency in heat exchange can increase the efficiency of cars, airplanes, and craft, thereby reducing CO_2 emissions. Among the many successful slams I researched, only two slams indicated that their research was needed for scientific reasons and not for aiding general society. Research on elementary particles in one slam was justified because we know too little about the universe, and the slam we investigated earlier on *The Voynich Manuscript* cited a lack of knowledge about the language in the book as justification for scientific research.

The second type of Science Slam I call '*Use of Science*'. These slams typically start with the slammer outlining a societal problem and end with the slammer explaining how science can resolve this problem. A slam that exemplified this was one which explored the problems associated with an ageing population, specifically having to mend elderly people's bodies when they fail. This slammer presented his research on generating implants for surgeons as an answer to this problem. A second example comes from a slammer who identified that bats have better teeth than humans and researched how this could be useful for humans because they have much more fragile teeth.

If we look at the content in the most successful slams in Germany, it becomes apparent that most slams provide advice which is applicable in everyday life. In the previously explored slam of Johannes Schildgen, the aim of the talk was to explain the advertising strategy used on online platforms like Amazon using the example of selling chips at a chip shop. The goal of his slam was to explain and understand why Amazon is so successful, and his task as a scientist was to improve the ways that product data is analysed. If we are to look at what percentage of the talk he spent talking about his own research, around 80% was spent on the content of his slam, for example the entertaining stories about Amazon's suggestion feature and the chip shop example and so on, and around 20% (the last two minutes) was spent on solution and evaluation, with a focus on his own research. In slams, it is viewed as favourable if you manage to use scientific knowledge and relate this or apply it to everyday society and make it accessible for a non-academic audience. The difference between 'Science and Use' and 'Use of Science' is that in the first type, 'Science and Use', societal use and application is an optional add-on; whereas, in the second type, 'Use of Science', societal use is the catalyst for the research in the first place. Both types, however, show that scientists are useful and that they can solve societal problems, be that helping people with health issues, improving everyday life, or saving the world.

Precarious Scientific Ideas and Supremacy in Science

I would now like to talk about two observations I made when researching the Science Slam. Firstly, I will explore the idea that scientists today seem to be forced to find new sources of capital. Secondly, I will discuss whether scientists see themselves as superior.

Science Slams can be an indication of the uncertain situation many scientists find themselves in today. It is common to see various sales strategies in Science Slams, which are in part, a form of self-marketing. These sales strategies are primarily a bid by the slammer to raise capital through selling their book, raising money for their project, or looking for sponsors for their research. Due to this, it has been suggested that there is always a commercial goal in slams.

A good example comes from a slam by André Lampe. Following his slam Lampe asks, 'Why do we need all of the things' (that he had argued for)? He answers this by explaining that interleukin one beta (the drug he is researching) is key in curing many diseases including Alzheimer's, Parkinson's, diabetes, leukaemia, HIV, hepatitis, and cachexia. Unfortunately, testing for these diseases by having regular blood tests is quite expensive (€10 per test) and can take up to 24 hours to process. Andre Lampe and his colleagues have invented a blood test that can be done very quickly. However, the test is still expensive because his team have to produce the antibodies for the test themselves. He used this as an opportunity to make an announcement, in which he took a sweeping step, swelled his chest up, looked at the audience and asked the industry to take over producing antibodies so the tests will become cheaper, which will benefit both Western and developing countries.[22]

The second observation regards supremacy in science. An example of this comes from a Science Slam that took place at Festsaal Kreuzberg in Berlin on 8 November 2011. It was organised by Haus der Wissenschaften and funded by the Federal Ministry for Education and Research. Researchers were asked to present in the field of 'Health Sciences'. This slam was the result of four preliminary competitions in Aachen, Braunschweig, Freiburg, and Leipzig. The top eight from these stages competed against each other at the final, which took place at Festsaal Kreuzberg, a venue used for concerts, readings, discussion evenings, parties, and fashion shows.[23]

This slam was by a slammer called Henning Beck, from Ulm, who researches the biology of nerve cells. He says that he finds pain and movement interesting because pain and movement are two processes that take place at the same time but in different locations in the body. He illustrated this in his slam by showing a photo of two boxers punching each other in a boxing fight, which he explained simultaneously. In the image on the left-hand side stands a white boxer who is shown to have 'a large brain', while on the left side stands a black boxer who apparently has 'a small brain'. Beck said that the distance from the motor centres in the brain to the actual point of a punch is more than five feet but this is bridged in a few milliseconds because the pulse travels at more than 400 km/h in the 'light-skinned person', but on the other hand, the black boxer's pain is slower because it is only about 50 km/h to the 'indeed pretty small brain'.

Beck continued explaining nerve impulses and how these relate to emotions and feelings, which are essentially electrical signals in our bodies. Beck argued that feelings can be processed very quickly, which Beck thinks is surprising because our nervous systems are incredibly complicated as we have billions of cells, a million-kilometre pathway, and nerves with unpronounceable names. Beck questioned how such a complicated system can function so efficiently and claims that one can answer this by talking about the nervous system. However, his research on the nervous system and the images used to illustrate the nervous system online generates mixed results. The image from Google Images showed too little information, the one from Wikipedia is also lacking but is a little better. Beck stated that he does not want to present the audience with a 'crappy kids' illustration' of the nervous system and so he showed the third image he found, which was from a specialist journal called *Apotheken Umschau*. Nevertheless, Beck claimed that this image is as 'false as everything else' because it shows gaps between the nerve cells and fibres, yet in reality these gaps are filled with helper cells or glial cells that are just as important as the actual nerve cells but are always forgotten in images.

Beck wore blue jeans and an elegant shirt. He acted very professionally, was well spoken and polite, although he spoke a little quickly. His presentation was not made with PowerPoint as is the norm, but with the software Prezi. He is a scientist primarily, but comes across as a knowledgeable polymath. Everyday images, such as inaccurate representations of the nervous system, amuse him. Most strikingly, in his talk we find a Darwinian 'survival of the fittest' emphasis. He argues that there are successful creatures and less successful creatures (some examples he uses include caterpillars versus humans and black boxer versus white boxer). His main thesis is that one can measure how complex a creature is via the speed of their nerve impulses[24] and thus he believes that humans are more biologically advanced than animals. As a 'superior' scientist, he seems to suggest that he is better at illustrations and more knowledgeable than the public. He additionally states that he delegates irksome work to his Greek student assistant, who no doubt he believes is less advanced than him.

Designing Slams Around Recipients and the Question of Translation

I will now summarise how slammers design their slams around their recipients, what kind of translation processes slammers use, and what I encountered at slams.

When slammers perform their slams, they usually think about how they can make their research relevant or interesting to the general public. Most slammers think about their prospective audience, where their slam will be located, and the possible reactions to their slam. For example, if a scientist thinks that their audience will not be 'scientific', will 'bore easily', or be 'misinformed', this will alter what they include in their slam. As such, prior to a performance, a slammer will think about their potential audience and try to imagine their performance taking place. The practice we explored earlier of watching old slams on YouTube is often used. However, although a slammer may imagine how a performance will go, there are bound to be some challenges when the slam is actually being performed. As such, one could conclude that the communicative process emerges when the imagined and the situated realities are brought together.

A slam in which the imagined performance was very different from the actual performance was Johannes Hinrich von Borstel's slam '*The Non-Christian Way to Almost Eternal Life . . . or . . . How Sex Will Save Your Life*'. Among other things, Borstel joked about the Pope's inability to control his urination, his delusions about being a representative of God, and his sexual abstinence that, according to von Borstel, has caused serious health problems like arteriosclerosis. He also characterised the Pope as of an 'advanced age', and, due to illness, 'recently unemployed'. His whole presentation was based around the idea that, in his imagination at least, the audience would find Pope jokes amusing. However, von Borstel was ill-prepared when he came to Münster—a very Catholic region of Germany—to perform his presentation. He quickly realised that people in Munster do not like Pope jokes and therefore the situated reality was far different from the imagined one he had created in his mind!

Borstel had a history of using inappropriate material and of being a bit of a taboo-breaker as we will see in the following example. At the beginning of his Science Slam career, Borstel ended his talk by giving some unusual medical advice to his audience. Although his style of presentation was quite formal, the content was risqué. Borstel gave a final piece of advice to his audience, which was that they should swallow semen as a home remedy for a sore throat. As expected, this made a few people in the audience laugh. According to Borstel, semen can help cure a sore throat because it can build a protective film of mucus and it is antibacterial. For this reason, he advised the audience 'bring oral sex to the bitter end'. Not everyone appreciated Borstel's advice, however, and some organisers advised Borstel to cut out this part of his presentation before attending the German championship, which he did.

Clearly Johannes Hinrich von Borstel did not learn from his mistakes because he got caught in a similar situation in a slam on animal testing. In some of his medical research, Borstel was required to do animal testing which he was planning to mention in his slam. One organiser warned him to be careful about mentioning animal testing when he was invited to speak at her event because from her experience animal testing was a big issue locally. Borstel ignored her warning but it did not seem to matter as no one complained about his talk and he was crowned winner of the slam.

Summary

In this section, I have tried to explain the **external structure** of the Science Slam. The Science Slam, as it exists today, usually consists of several key factors or features: financial dependency, organisation, a typical social grouping, and certain rules of communication. We have learnt that financial support can come from a variety of sources; organisers are usually either employed by conferences and museums or funded by sponsors. Common sponsors are the local city, local university, banks, insurances companies, and magazines. Organisers can usually be divided into two groups—science school dropouts who have gone into business themselves or science communicators who work for an organisation that specialises in science. As we have explored, the typical social group at slams consists of those who are young and sophisticated. Science Slams take place, for the most part, in locations like theatres, rock concert venues, and clubs, while loud, youthful music is usually played and drinking alcohol is common. All slams tend to share a certain rancid charm which is perhaps due to the fact that slams take place in crowded places, are relatively cheap to run, and an informal atmosphere is encouraged.

The **inner structure** of the Science Slam consists of an aversion to professional jargon, with a preference for informal, everyday vernacular instead. Having a dialect or accent is not considered problematic in slams. On the contrary, many successful slammers emphasise their dialect. In terms of the slam's length, slams should be short. This results in many speakers talking rather quickly and having a continuous flow of speech, rather than taking time to pause. Code switching is another key feature of the genre's inner structure, whereby slammers switch between professional jargon and everyday language, and back again. The use, and presentation of, physical, scientific objects is common, as is the use of multi-media equipment, which can include cameras, projectors, microphones, and laser pointers. Most presentations have a triadic structure which exists between the presenter, their audience, and the objectivations. Presenters offer their bodies (Knoblauch 2005b), their body language, their voices, their clothing, and their attitudes as a representation of what science and scientists are like. The presenter–audience relationship typically includes acts of presenter–machine interaction, which are usually well orchestrated and timed. Nevertheless, the genre generally supports an 'in-the-making' aesthetic, whereby the audience sees the event being prepared as they wait.

Science Slam presentations usually have a standard order or sequence which consists of four parts. It begins with an opening sequence (a) wherein the moderator introduces the presenter, who then comes onstage and prepares to begin their talk. The second part is called the orientation sequence (b). In this, the slammer outlines their research, which usually fits into one of the following three topics: scientific thinking, a scientific object, or a societal problem. Thirdly, is the solution sequence (c), in which an ambiguity is resolved, or the role of the presenter is clarified. Slams end with the evaluating sequence (d), in which the presenter reflects on their research. In this sequence, it is not uncommon for the speaker to contextualise their research in a wider scientific context or to demonstrate what their research contributes to society.

Finally, is **situated realisation,** which is dependent on the communication between moderators, scientists, audiences, and their objectivations. As we have seen, a slam's order or sequence is determined by the moderator. The situated realisation of a Science Slam can be characterised as 'humorous science communication', and the expressions and emotions of the audience are a part of this interactional organisation because speakers tend to present themselves as humorous and they also allow short breaks for laughter or interjections from the audience.

Communicative constructivism as social theory of contemporary society (and further development of social constructivism) has to develop a stance on how new communicative genres emerge, consolidate, and change through communicative actions. A theory about genre must take into account that anything other than linguistic objectivations (such as visual jokes on PowerPoints) can become a characteristic of the genre. In this chapter, I have partly tried to extend my analysis of genre and, in this way, widen the focus. It is worth nothing that in my study, physical objects and visualisations were part of the internal structure. In the external structure, the venue was included as a relevant theme due to its status as an institutionalised place with a history and atmosphere. In the situated realisation, meanwhile, I tried to include more focus on aspects of embodied communication.

In Chapter 5 I have described how the communicative genre of the Science Slam has evolved and changed over time. A key part of this is the change to the characteristics of the genre which I have illustrated empirically by using material from interviews, newspaper articles, and pictures. I have also attempted to describe the current realisation of the genre based on video analysis and ethnography. I have empirically analysed the Science Slam field in close detail and have characterised it in various aspects. I would now like to consider the methodological condition of theorising by reflecting on a theory of the genre for a future development in the perspective of communicative constructivism.

In my research, the exploration of the relationship between social networks and social movements and the merging of online and offline practices has become central. In Chapter 5, I have shown how social networks have become platforms for the mobilisation of the Science Slam movement through the creation of an archive of successful slams on YouTube. This has allowed for the diversification of informal learning and the communication culture, as well as allowing for the offline and online worlds to merge. This sharing of information has manifested in various ways; through the use of images, videos, and jokes and also through scientific personas, references, and translating processes. In this way, the translocation of social space (Knoblauch 2017, 341) has become important. The face-to-face, in the flesh situation transcends and becomes mediatised when the performance becomes part of the technical knowledge archive online. Communicative actions are, therefore, permanently present for anyone who has access to the internet (ibid.) and can be visited by many people (including those who are planning to perform) as many times as they like.

On this note, one could argue that the development of the Science Slam genre is not only due to the actual 'live' slam and the gathering of people for this, but also due to the informal learning practice of watching YouTube videos of slams that have already taken place. Many people gain their subjective knowledge about the genre based on situated communicative experiences, and through digital environments this is mediatised. In such informal learning environments people use, what I call, creative modification. In this, science slammers use features from successful slams without providing citation, images from the internet without thinking about copyright, and reproduce digital media content but re-invent the content in their own way. The various changes to the subjective knowledge, situated communicative action, and intersubjective shared knowledge about science communication happen as a result of the specific online and offline experiences I have documented.

Notes

1 At this time (February 2012), Giulia Enders was in the middle of writing her dissertation on Bacterium (specifically Acinetobacter baumannii) at the Institute of Microbiology, Goethe University. I learned, from a newspaper article, that she wanted to become a Gastroenterologist and she hoped that there would be more research into the positive influence probiotic bacteria have on people's wellbeing.
2 Translated from German, accessed September 25, 2015, www.faz.net/aktuell/wissen/darm-mit-charme-von-giulia-enders-12891555.html.
3 Accessed September 25, 2015, www.youtube.com/watch?v=2qo3ueVlyUY.
4 A few commonly used but contrasting locations include Festsaal Kreuzberg (Berlin), Grammatikoff (Duisburg), Kampnagel (Hamburg), Bogen 2 (Köln), Jugendhaus KOMMA (Esslingen), Jovel Music Hall (Münster), Lido (Berlin).
5 Examples of this include the Max-Planck Institute of Physics, the Science Centre of Berlin, and the IdeenExpo.

6 'So, you do not make money with it. With every slam there are a few hundred Euros left but then you have to buy material or a new presenter or things like that. So, I see it more as a hobby. A hobby with benefits, but a hobby' (MS#49).

7 I think the funding from the Ministry of Education, at the very least, has had an impact on the development of the Science Slam and, therefore, which researchers have participated at events.

8 The question of milieu was also addressed in visitors' interview statements before one slam. Most of the interviewed visitors expected the audience to be composed of mainly young and sophisticated people. Older visitors or those with different clothing styles mentioned that they felt out of place. A 32-year-old nurse said, 'we just found out that our outfits are inappropriate. . . . The ZKM has this certain level and it is just this special clientele'. A 48-year-old teacher said 'I don't know if we find more young people and I am out of the ordinary, I am curious'. And a retired 68-year-old man said that he hoped the young people would not try to impress the audience by using slang words.

9 My thoughts on the audience at the German championships in the Festsaal Kreuzberg were as follows. 'The audience seems young. I would guess they are all around the age of 22. If I had to guess where they came from I would think they are students of the Technical University of Berlin. I would suggest many of them are Engineering students. The audience are mainly dressed in blue jeans and black t-shirts. I see a lot of people with glasses. Lots of people are talking in the audience. Every other person has a beer, including the slammers' (MH#22).

10 The official German Final is self-funded and organised by a group of Science Slam organisers.

11 The date of 2008 was probably chosen because the event was organised by the Haus der Wissenschaften. People from the Haus der Wissenschaften have their own innovative narrative.

12 I once attended a Science Slam where the speaker was interrupted by a member of the audience—a guy screamed 'You are talking bullshit!!' At the end of the slam the moderator condoned the interruption and said that no-one should interrupt the speaker in such a way.

13 The competing scientists were Dominikus Krüger (Informatics, University of Ulm), Johannes Schildgen (Informatics, University of Kaiserslautern), Joscha Beninde (Bio-geography, University of Trier), Matthias Stahnke (Aerospace Technology, University of Aachen), Oliver Marchand (Alumni, ETH Zürich), Sascha Vogel (Physics, Goethe-University), Siegfried Bolek (Biology, University of Ulm).

14 The competing scientists were Sarah Jarvis (Bernstein Center Freiburg), Mohamed Zidan (Universität Mainz), André Lampe (Freie Universität Berlin), Henning Beck (Universität Ulm), and Felix Büsching (Technische Universität Braunschweig).

15 Accessed September 25, 2015, https://www.youtube.com/watch?v=xm5bcUFqa30 The participants had ten minutes to present their research. The moderator (Roland Kremer) introduced each speaker and finished up when the talks were over.

16 Accessed September 25, 2015, https://www.youtube.com/watch?v=WVvqqxTjaI8

17 AJ is also the name of a member of the band 'The Backstreet Boys'!

18 I once witnessed a moderator being critical of a 'handsome and smooth' presenter who made no errors. In the slam community, it is widely believed that slams are not the place for perfect people, but more suited to those who are imperfect. This guy did not win the slam and I never found out why.

19 Accessed September 25, 2015, https://www.youtube.com/watch?v=v-9SSg4DpcI

20 The slam was organised by Julia Offe and supported by Geo Magazine and the job website 'academics.de'.

21 The competing scientists were Mogan Ramesh (Process Engineering), Reinhard Rem-fort (Quantum Physics), Kathrin Steyer (Genetics), Lydia Möcklinghoff (Biology),

Johannes Hinrich von Borstel (Medicine), Sarah Kohler (Communication Studies), and Christoph Seibel (Physics).

22 Lampe further explained that some astronauts suffer from cachexia and so the European Space Agency supported him through his diploma. As the industry has funded his past projects, he searches for more support for his new project. Lampe tells the audience that if they are asked why we need space research, then the audience could answer that space research supports a test that saves time and money for hospital patients. We can conclude that Lampe offers legitimisation for space research.

23 The participants were Vinay Rambal (Charité, Berlin), Helga Hofmann-Sieber (Heinrich-Pette-Institut, Hamburg), Patrick Seelheim (Westfälische Wilhelms Universität, Münster), Sarah Jarvis (Bernstein Center, Freiburg), Mohamed Zidan (Universität Mainz), André Lampe (Freie Universität Berlin), Henning Beck (Universität Ulm), Felix Büsching (Technische Universität Braunschweig). Other people involved were Dr. Helge Braun (Bundesministerium für Bildung und Forschung), Roland Kremer (Moderator), Britta Eisenbarth, and Markus Weißkopf (Haus der Wissenschaft Braunschweig).

24 According to Beck, consuming animal meat changes the intelligence and complexity in evolution. Beck argues that consuming meat is important for the development of the human brain and that it leads to higher levels of intelligence in humans.

7   Science Slam in Contemporary Society

The New Art of Old Public Science Communication

As we have learnt throughout this book, the aim of sharing science with the public is not new. The history of science and its relationship with the public reveals that there has always been a struggle in the sharing of knowledge between scientific experts and non-experts. The context in which this relationship began was in the ideas of a sophisticated public bourgeois, at which time chambers of curiosities, museums of natural science, zoos, botanical gardens, observatories, and aquariums became popular. Since then, the popularisation of science and the education of the public has been an aim of Western societies. Instead of listening to wandering orators (Wanderredner), people today, in Germany at least, attend Science Slams. The public want to 'be entertained by scientific topics' and 'get excited about research'.

After the Second World War and the educational expansion of the 1960s and 1970s, when science became differentiated and specialised, the public became more critical about the separation that existed between themselves and science. In response to this, several new public science communication events emerged, particularly observable from the 1980s onwards. As one of these events, the Science Slam shows that new ways of legitimising science have been established in the field of science communication. The Science Slam demonstrates a dissolution of boundaries in the public sphere.

Habermas's judgements about the public were related to the question of how communication genres do or do not empower a public. For Habermas (1990) the end of the broad public discourse happened before the digital age. According to Habermas, the bourgeois and literate public disintegrated at the end of the 19th century. The subsequent post-bourgeois, mass media influenced public was staged, one sided, uncritical, and undemocratic. For him, this signalled cultural decay. I did not plan to agree or disagree with Habermas's judgement on the public in this book but wanted instead to offer an empirical description of a type of public science communication. However, Arendt's (1960) description of the birth of totalitarianism completely changed my perspective on current developments and the formation of the public sphere. According to Arendt, power arises in the common actions of people in public spaces. The public sphere lives due to the plurality of

DOI: 10.4324/9781003172635-7

its participants, who in dialogue with each other, create an awareness of a multi-perspective world. This is especially due to their diversity and antagonism. If we want to understand public spheres as socially and communicatively constructed spaces in which different perspectives and identities relate to, and connect with, each other (Fraser 1992; Arendt 1970a), cultural struggles and political debates are relevant. Through extending deliberative models (cf. Habermas 1990), I have analysed the public sphere as an empirical category. The public sphere is neither normatively postulated nor ontologically presupposed, but communicatively constructed. I agreed with Fraser (1992) in her critique of Habermas's theory when she asked for alternative counter publics. I focused on the processes that occur through scientists' communication with the public and, therefore, in this project I considered how the boundary between science and the public is enacted and performed in Science Slam performances. When considering new genres like the Science Slam, I asked in what way this communication contains an asymmetry of science communication that produces or continues dissimilarity.

According to Fraser, identity formation constitutes a significant part of public participation. In this context, the analysis of social identities in the past, as described by historians like Shapin, must be considered in a different way. In Chapter 2, I gave an overview of the history of repression of women in science. I explored new forms of the scientific persona which have arisen, and I observed how these have all been present in Science Slams. The gentleman from early modern times, who was trusted to speak the truth, has been replaced by new types of scientists. I noted that the public seems to be interested in the creation of scientific knowledge, and in acknowledging the human side of scientists today, which humanises the labour of science and results in scientists exposing their knowledge, feelings, thoughts, and perceptions. This phenomenon could be referred to as an 'exhibition of scientists'. Examples of the new types of scientist are the 'human man', the 'popular culture man', the 'working man', the 'non-pseudo man', the 'entrepreneurial man', and the 'ascendency man'. These are just a few types of the new scientific personas that have emerged in this rich field in recent years. These shifting identities can be compared to Bröckling's (2007) concept of an 'entrepreneurial self'.

If we believe Fraser, science communication must be acknowledged as a specific cultural setting, with certain rules, which includes some people and excludes others. It is important to bear in mind that there are multiple, coexisting public spheres in an egalitarian and multicultural society which reflect the differences in such a society. Fraser's thoughts on subaltern counterpublics have consequences for perceptions of democracy and political equality, and they also serve as a stark reminder that the communication of science is not open to everyone because societal inequalities prevent certain groups from becoming part of the discussion.

This is unfortunately the case with Science Slams because certain parts of the population are excluded from slams. This is, in fact, one critique of the Science Slam by poetry slammers, who claim it is a bourgeois style of communication. It could be argued that there is nothing revolutionary in building arenas for debate for people who are already privileged, since the genre represents a specific relationship between science and the public, and is also part of a particular milieu that is present in places of popular entertainment. The milieu of the Science Slam could be described as a form of urban knowledge with a scientific, positivist orientation; a self-actualised milieu with the aim of personal self-realisation which is very much oriented towards excitement, and is focused on high culture. In Poetry Slams, pretty much anyone is allowed to go on stage as a participant. In Science Slams it is only members of the scientific field who can go on stage and present.[1] In the beginning, the Science Slam might have been considered to be a form of subaltern counterpublics since non-scientific knowledge was allowed to be presented. In 2016, however, the Science Slam could only be considered subaltern counterpublics in relation to the institutional field of science.

Boundaries between science and the public can be enacted by socio-technical arrangements, orders of communication, or visual representations. In public science communication, the questions of 'what is public?' and 'what is science?' are shaped by communicative processes. I observed the enactment of contemporary science communication and highlighted current socio-material arrangements and interaction orders to further identify the new genre. In focusing on the Science Slam, in particular, my project aimed to answer the question of how publics are communicatively constructed.

In the next chapter I defined some of the generic features of Science Slam presentations and provided empirical evidence of these features. I also highlighted current socio-material arrangements and interactions, and explained how the Science Slam is both linked to and different from previous forms of scientific communication.[2] While empirically studying the communicative genre of the Science Slam I observed particular changes of situated communicative action. This book has explored and documented humorous interactions between scientists and audiences; a loosening up of power structures in comparison to classic lecture formats; and insights into the significance of the quality of the presenter's orchestration, including the importance of timing. The book has further seen evidence of scientists offering their unpolished bodies in order to effectively communicate science and how scientists use visualisations (like scientific images), translation slides, physical objects, an everyday linguistic register and code switching to communicate. In addition, it has explored the 'rough-diamond' style of communication and how the presentation of science can be split into two categories—'use and science' and 'science of use'—and how science slammers evaluate previous Science Slams and sometimes creatively modify them.

In this book, I have clarified how scientists describe their often-problematic relationship with the public and how they embody these relations. As I learnt from my interviews with slammers and organisers, science slammers predominantly leave out methods, statistics, and theory. These omissions sometimes make scientific research more comprehensible, but this is done at the cost of empowering the audience to understand scientific procedures in greater detail. Often, presentations include strong asymmetries with the speaker and audience. This is shown in the suggestion that slammers explain a topic in a 'very primitive' way, in barefaced ways of confronting the audience with context-free facts, and also in examples of visual and performative asymmetries. By leaving out methods, statistics, theory, and diversity, Science Slams do not allow audiences a first-hand understanding of science in action. However, since other aspects of scientific life, such as the reference to the workplace, the everyday activities of science, and the scientist's self are central to Science Slams, listeners are provided with a second-hand access to science in action. In a similar way to feminist analyses of STS, the Science Slam contains both 'invisible authors' of scientific knowledge (who leave out methods), and also 'visible authors of knowledge.'

Are Science Slams political? With the feminist perspective on STS in mind I studied gender inequality in science communication empirically. Studying language, performance, and the communicative context helped me to understand why female researchers are still underrepresented in science communication. Science slams are an example of how popular scientists position themselves in science communication events. Chapter 6 explored how Giulia Enders, a female scientist from the discipline of medicine, became a successful science slammer. She has had a significant impact on scientist's public image, and her slam also marked my own entry into the world of Science Slams. She is an example of how traditional ideas about science are challenged in Science Slam presentations. Though Enders is not indicative of a 'Great Woman Theory' any more than the Science Slam founders are, her success has been influential and this has had an impact on herself personally, as well as on novel, public legitimisation practices, science presentation, and interactional organisation. I included her example—unusual in some ways when compared to the current milieu of science slammers—to encourage the hope that such public science communication can have an impact on breaking down the conventions of science communication. Although the success of Giulia Enders' presentation might lead us to hope for new norms of gender, technical jargon, and visual practices in scientific communication, these hopes are not entirely fulfilled by the Science Slam. Even though the technical jargon and visual practices of slammers in Science Slams are different from those used in university lectures, the marginalisation of women and other social groups still remains an issue.

Empirical observations in situated science slam settings have shown that, even though the Science Slam enables revisionist representations of science for men, women remain silent and invisible. This re-narrates classical feminist theories. Simone De Beauvoir ([1949] 1974) described how difficult it is for women to develop an independent and free subject status, beside the societal ascription as an object in which they passively have to develop an ego in relation to men. Drawing on typical tropes from patriarchal societies, in some Science Slams women are presented as objects of desire, as hardworking assistants, or as aunts, mothers, and grandmothers. To interpret these images performatively, we must look at what kind of identities are produced through their invocation, and in what way these allow for presenters to connect with their audience. The Science Slam could be understood as an expression of the still-problematic gender inequality in science (especially in natural, applied, formal, and health sciences) and wider society. Additionally, slams often include offensive and marginalising representations of minority groups. Organisers of Science Slams wish for slams to be diverse, but in my research I came across many typical tropes from patriarchal societies. These representations should be understood as a reflection of social power relations, rather than as a product of the genre.

Although it is problematic that such discourses are displayed on stage, scientists allow themselves and their patriarchal discursive moves to be criticised by situated audiences, other scholars, policy makers, and the media. As with information on general PowerPoint presentations (Schnettler and Knoblauch 2007, 19), knowledge becomes more tangible in Science Slams. This is reinforced by the practice of organisers uploading slam presentations onto the internet, and their recognition of slams as a media product. In this way, the few women who were on stage at the time of my survey can also have a major impact. Social media makes scientists more visible, but whether this is a good or bad thing is up for debate. Either way, the Science Slam has become a positioning practice of science. Slammers embody and localise knowledge claims. In contrast to the typical unlocatable knowledge claims of modern scientific research, slammers 'get naked' in public, figuratively speaking, when they present their slams.

The public credibility of science and the trust in institutions is not a solely intellectual process, but based on material social relationships (c.f. Wynne 1992, 281).

> Credibility and trust are themselves analytical artefacts which represent underlying tacit processes of social identity negotiation, involving senses of involuntary dependency on some group, and provisional or conditional identification with others, in an endemically fluid and incomplete historical process.
>
> (ibid., 299)

It is very important that scientist respect specialist knowledge of others, that they communicate their uncertainty, and mention the fact that knowledge can always change. This is necessary for sustainable science communication. Arrogant, too certain, and uniformed styles of communication can lead to alternative constructions of reality. Conspiracy theories are often not only based on intellectual mistakes of people, but also on social failures of public communication experts (ibid., 293).

> The best explanatory concepts for understanding public responses to scientific knowledge and advice are not trust and credibility per se, but the social relationships, networks and identities from which these are derived.
>
> (Wynne 1992, 282)

If we want to communicate science, we not only have to explain how science produces knowledge, but also build social relationships. The 'supremacy of science' as a communication strategy in science slams that I have observed is more dangerous than one would first think. Maybe scientists are trying to move towards the 'truth'. On the way there, however, they are often wrong—true science communication must include this part of the story. Untrue knowledge does not always have its source in social interests or cultural values, but can also be based on other sources of interference.

Conversely, scientific knowledge is not the pure result of social interest, or of cultural values. Unlike conspiracy theorists, actual scientists are challenged by the 'real world'. They cannot simply upload a video or paper onto the internet claiming that the earth is flat without any knowledge or research to back this up. They have to acquire an academic habitus, prove they are eager to learn in a responsible, knowledge-oriented, institutional field and demonstrate that they understand established stocks of knowledge in order to become university staff. When they find new knowledge through scientific methods and research, they encounter another barrier in the form of resistance to the science or scientific discovery. Following Berger and Luckmann (1967), I encourage a dialectical process describing the relationship between knowledge and its social base. In this dialectic process, the expression of social identity plays a crucial role. Scientist and other people are social groups 'attempting to express and defend its social identity' (Wynne 1992, 297).

The backgrounds of scientific knowledge and conspiracy theories are quite different, so it is even more of a challenge when people who have academic credentials start to accept and spread conspiracy theories. On the other hand, both can create performativity when people in society trust in them. The competition between the two can probably only be won if scientists step into the public eye more often, and prove that they are more trustworthy and that they describe reality more effectively than conspiracy theorists. Based on sociological theory and my empirical results, I have argued that the Science Slam has the potential to regain the public's trust in science within the context of post-truth politics. Indeed, the Science Slam has been described as a potential 'access point' between science and the public in a knowledge-based society. Shapin (1994) suggested that scientists must convince others in social processes to get them to believe in knowledge. Therefore, I looked at the interaction and the performance of these processes of conviction. Shapin further argued that next to trust in systems, the patterns of traditional trust still exist today. This has been confirmed by my research. The self of the scientist becomes more visible in, and through, Science Slams. Historically, the scientific gentleman was trusted to deliver truths; today we can observe new scientific personas, such as the 'popular culture man' who is also a legitimate author of knowledge.[3] One of the features of the 'popular culture man', for example, is the staging of a popularised self, whereby scientific activities are explained through popular culture. Techniques of re-enactment, adoption, citation, and variations of former presentations are also common. Since scientific knowledge cannot be considered, per se, as separate from everyday life, it has to be presented like all other forms of knowledge and this performance relies on forms of communication that are also common outside of science. Slammers could be considered creative modifiers of cultural repertoires, through their adoption of comedy and popular media. This is not a hybridisation of different styles of communication, but a new form of scientific communication, which is strongly related to everyday practices and spreads interdisciplinary science communication as well (Wilke and Hill 2019).

Motives for the Establishment of the Science Slam

The construction of the public and the construction of science can be studied by researching Science Slams. Slams also reveal scientific (re-) presentations of overall needs in society. In Chapter 5 I explored the perspectives of actors in the field. Based on interviews with them, together with the ethno-theories from the field and newspaper articles, I reconstructed the story of how the Science Slam was established. Processes of innovation often begin with the question of whether certain norms or standards are still appreciated and relevant. Events like the Science Slam review potential communication problems and try to establish new forms of justification. From a theoretical perspective, institutional structure has to be founded by normative and cognitive legitimacies. Therefore, I highlighted the specific expectations, norms, and values of Science Slam participants. I explained why people who are involved in Science Slams see a need for their event and what their experiences participating at the event are.

Communicative genres have a socially accepted explanation of how genres come into being. In a similar way, there are certain expectations as to how Science Slams will run and be organised. Genres must often deal with certain communicative problems and so, too, does the Science Slam. The legitimisation of the genre points to the communication problems of science. The protagonists of the Science Slam present a kind of ethno-theory of science communication that has striking parallels to the critique formulated in PUS. I will now sum up the various issues that organisers and slammers encounter and expand on these.

If we look at the basic communicative task of the Science Slam, it is to acknowledge and validate science in public. Talks have to deal with the basic communicative problem of differentiated societies in order to express how scientists' research is relevant for society. This addresses the communicative problem of presenting a scientific topic and making it understandable and relevant for the audience. Scientists aim to show that their research is valuable and in order to do this they must make their research visible and try to receive positive attention for it. Science Slams present a new way of legitimising science in the field of science communication. At the same time, the Science Slam is an institutionalised genre of communicative action. Institutionalised orientation for actions—such as time limits, competition rules, a need to present self-made scientific content, a need to translate them, and a requirement to create emotional responses and establish an atmosphere—set the frame for these new forms of legitimising science.

Table 7.1 Different Aspects of the Science Slam and Problems With Them

Subject	Problems Organisers Identify With This Subject
Knowledge Communication:	Organisers claim that there is a growing need to communicate with people who are not scientists. In their view most university faculties still see little value in such communication. The belief in the value of disseminating scientific knowledge is rooted in a commitment to scientific authors, and this belief is confirmed by the fact that Science Slam organisers worry whether knowledge in slams is original. It is important that slammers are the authors of the knowledge they are presenting.
Making Science Entertaining or 'sciencetainment':	The main reason people attend slams is because they want to be entertained by science. The act of making science entertaining is seen as necessary to allow a non-scientific person to understand complex science. Organisers call for traditional tensions to be broken so relationships between slammers and audiences can be established.
Placing Knowledge:	The Science Slam can be seen to support local community infrastructures. Organisers and slam moderators alike emphasise that locality is key. Meanwhile, the dialogue used in slams is often viewed as having been locally produced, while slammers are seen as regional representatives.
Knowledge Legitimacy:	Organisers argue that scientists have a responsibility to share their research with the public because the public pay for a large amount of scientific research through tax.
Experience (as a Product):	Organisers believe that the public are interested in scientists and so there is a need to humanise science. Slammers are encouraged to do this by showing their feelings, thoughts, and perceptions when presenting slams. This allows experience to be represented, produced, and exchanged.
Competition, as a Form of Empowerment:	Organisers believe that the 'slam' nature of Science Slams is crucial. Organisers contend that people enjoy judging others, and so the public are much more likely to attend talks when they are allowed to judge them afterwards. Organisers also believe that there is high discursive value in the audience sharing and discussing their thoughts and judgements.
Learning:	One of the goals of the Science Slam is to get people excited about research. More specifically, the Science Slam is viewed as an opportunity to convey an important message to the public, which the audience takes in subconsciously, whilst also taking advantage of 'sciencetainment'. Generally, an explicit educational mandate is not pursued and seen instead as more of a voluntary addition.

Subject	Problems Organisers Identify With This Subject
Empowering the Audience:	The ability to judge slams is considered to be an act of empowerment. At the end of a slam, the audience usually votes on whether the content of the slam has been communicated effectively. The audience thereby has direct contact with the speaker through acoustic (emotional) responses.
Platform for Novelty:	The Science Slam community has a broad understanding of novelty, and novelty appears in two forms in the Science Slam. Firstly, novelty appears in the form of new research that is presented. Secondly, novelty appears in the freedom to choose between various technologies and creatively perform a presentation.
Interdisciplinary Communication:	The Science Slam can be seen as a platform for both scientific/non-scientific communication, and interdisciplinary communication. Organisers believe that diversity among both audiences and participants is good and, therefore, any over or under representation is viewed as problematic.

The Visual Order of Legitimacy[4]

In the introduction of the book, I raised the question if a shift in the form of communication (from text-based to visual-based communication) might have consequences for dealing with knowledge itself. In the sociology of science debate about the visual, it has recently been stated that in consequence of reflexive scientification and the imperative of interdisciplinary and transdisciplinary, the impact of the visual in science has grown (Beck 2013). My empirical data has shown this to be true. In terms of methodology, I came to this conclusion not by concentrating singularly on technologies of visualisation, visual 'content' or institutionalised communicative practices, but by bringing these three factors together. By analysing the technological aspects, I recognise the durable technological institutionalisation of science communication; by considering the visuals themselves, it becomes clear that their use is only loosely tied to more traditional forms of science visualisation (e.g., in the diagram or table); and by analysing communicative practices, the embodiment of visualisation and their institutionalised relevance was confirmed. This triangulation of methods enabled the analysis of different conditions and practices, which—working together—stabilise the transformation of science communication on three levels, which are closely intertwined, making it unlikely to disappear anytime soon.

Thus, I have been able to show that scientists today increasingly depend on visual practices in developing new knowledge and communicating it to others. Empirical studies on communicative practices in a contemporary knowledge-based society have shown that the increased need to communicate with many others has resulted in 'the ubiquity of PowerPoint' (Knoblauch 2014). I would like to add that the ubiquity of visuals follows a similar path (Traue 2013). Contemporary communication culture creates an order that is no longer based on faith in substantial truth but rather replaces written texts with visual patterns to get in touch with others, to convince them, and to gain legitimacy for one's own standpoint (re-presentational knowledge). As visual conventions grow, we find that legitimisations are to a much greater extent performatively produced in communicative processes. The more central communicative work becomes, the more central the staging of the self (Soeffner 2001).

Against the background of this presentational duty (Schnettler and Knoblauch 2007, 270), the whole body is progressively replacing the complacent staging of an omniscient voice from the 'ivory tower'. Instead, a performative self has taken centre-stage. In the final analysis, it is this personal self which has to cope with the disadvantages of the new forms of knowledge communication and the empowerment of the audience. I aimed to stress the communicative and translational character of the visual in science and looked at the use of visuals in transdisciplinary science communication as part of communicative action. Using the example of Science Slams, I demonstrated that visuals of scientific knowledge are used as translational tools, as the genre focuses on digital slides and visuals. I frequently observed the use of everyday images as well as scientific visuals. The refusal or inability to work with visuals is seen as a major reason for the failure of humanities to gain audiences in comparable formats. The idea of visualising things has therefore been deeply internalised by the actors of Science Slams. Not only opening arguments, research questions, and findings are regularly visualised, but also the processes of collecting, preparing, and analysing data are often illustrated by visuals (or a series of visuals) and presented with laptops, laser pointers, and presentation software. In consequence, presenters in Science Slams have to performatively orchestrate their shows in space and time. The presenter's body is central not only as a speaking machine, but as a major tool to focus the audience's attention through the prefigured audio-visual objectified externalisation of research results. Alongside its orchestrating role, the body and its visual reflection move into the centre of attention.

From this point of view, a lack of re-presentational knowledge is clearly a handicap in transdisciplinary contexts. An inability to communicate with appropriate visual representations can easily lead to a lack of recognition. My empirical examples show an increased need for presentation in modern knowledge and science communication. The requirement for scientists to possess presentation skills has increased enormously in the context of transdisciplinary communication. In many disciplines, therefore, a form of presentational knowledge has developed, which must be regarded as essential for the institutional visibility of scientists and scientific institutions themselves. In my examples, the success of scientists in Science Slams is clearly tied to visual literacy and the ability to design their re-presentations in a way that meets the needs of the addressees. Language and visualisation must be tailored to the recipient in order to be heard and perceived as part of the scientific canon.

As Berger and Luckmann (1967) have described, scientists used to underline their authority with age-old symbols of power and esoteric signs, which included outlandish costumes and an incomprehensible language. A certain body of professional knowledge or legitimating machinery was at work and empowered scientists to distinguish scientific proof from quackery. Today, scientists in transdisciplinary contexts need to possess creative and artistic skills to underline their authority. Visuals are part of the scientist capital. In the past STS researchers have pointed to the question of how the power of science may relate to the visual tricks through which science is embodied. The circle of credibility (Latour and Woolgar [1979] 1986) that scientists use to gain acceptance was described as part of a gigantic scientific instrument, a panopticon, which allows scientists to produce visual consistency and dominate others. If today Western scientists with non-visual representational skills have to expect a loss of power, I conclude that the scientific panopticon has been upgraded to include artistic and popularised devices. The art of convincing others today not only relies on an outlandish costume and an incomprehensible language, but on weapons from art, popular culture, and economy. Haraway once opposed the 'invisible conspiracy of masculine scientists and philosophers' and the 'embodied others, who are not allowed not to have a body, a finite point of view' (Haraway 1988, 575). Disembodiment and universal claims were described as part of the Western scientist's bag of 'God-tricks' and as a view from nowhere. Today scientists still use the 'God-trick' to uphold power, but in transdisciplinary contexts they additionally position themselves in performances of multimodal communicative actions. They increase the visibility of their science and their selves. As we have seen, the interaction order in science has changed in genres like the Science Slam. In situated settings of science communication, new interaction orders are loosening up old power structures and strengthening new ones, within which disciplinary 'God-tricks' are increasingly supplemented with communicative visual practices.

The Economisation of All Public Spheres

The post-democracy debate (Crouch 2004) argues that in our current contextual situation the economy is constantly growing, and politics fails to keep up with decision-making processes (which should be slow and thorough). This paradoxically leads to a situation where there is a growing need for experts in politics, who outsource decision-making processes and legitimise political decisions, which are then uncritically accepted in parliament. Elites stand to profit from this system where politics represents economic interests, rather than what is best for the people. According to Crouch, this has led to an increased apathy towards political matters among citizens. We find reference to this idea in Hannah Arendt's work.

In discussions about post-truth or post-democracy (Crouch 2004) scholars wonder if neoliberal thought patterns and arguments undermine all social and political institutions, and therefore have a hegemonic status. If this is true, we should find an efficiency orientation within scientific research and communication. In the so called 'Mode 2 of Knowledge Production' there was a shift towards interdisciplinary and transdisciplinary research in science. Science became more open to external influences outside its discipline (especially the influence of those who have money), and scientific knowledge had to become more robust so it could stand up to certain markets and to society. As we learnt, this resembles discussions that Science Slam organisers have had about the importance of communicating with the public.[5] Organisers of slams still argue that it is the scientists' responsibility to communicate effectively with the public. Within this perspective, the Science Slam could be seen as 'Mode 2' par excellence, because it takes the performative pressure of 'academic capitalism' (Münch 2013), as well as 'entrepreneurial science' (Etzkowitz 1998) and 'neoliberal competition' (Beer 2016) and turns these into a form of science communication. In the Science Slam, knowledge is asked to be heterogenic, non-hierarchical, transdisciplinary, and above all, useful for society. It is a platform for expert and non-expert communication, for interdisciplinary exchange, and also provides an opportunity for scientists to find sponsors and connect.[6]

The Science Slam is a form of city marketing, as it supports regional and local scientific and economic infrastructures. In this view, scientists are 'entrepreneurs' or 'starry-eyed idealists', who communicate their new knowledge with the new world. The neoliberal self and economic ways of thinking are present in many of the modern scientific personas that I found (entrepreneurs, engineers, men striving for success, men adding value). This is comparable to Daston and Galisons' (2007) concept of a new scientific persona, who combines the ethos of the late-20th century scientist, with the technological focus of the industrial engineer, and the artistic ambition of an artist, which was explored earlier. I also observed the rise of 'popular science' and the 'popular self' in slams. The term 'popular science' is an indication that communicative practices, analysed in the context of Science Slams, have spread within society. Participants of Science Slams act as if references to popular culture and the economy are central, integrating features of science communication within contemporary society. This form of economisation and popularisation of science can be interpreted as an indication that the historically humanistic tradition and rhetoric has retreated (Daum 1998, 53), while natural science, business or utilitarian thinking, materialism, and the commercialisation of society are no longer stigmatised.

Similarly, the emergence of the Science Slam correlates with the expansion of public funding for science, and the increasing orientation towards third-party funding. Correspondingly, I found that the public within Science Slams are economically framed. As a form of 'sciencetainment', Science Slam organisers sell 'scientific experiences' and 'informal learning' as a product to the public. Slammers are indirectly aware that they need to produce an experience for the audience, and their own life experiences and scientific research are payment for this labour. A criticism of this is that the competitive nature of neoliberal societies manifests itself within the competitive aspect of slams. Weber would call this a capitalist university enterprise (Weber 1977). Reinforcing this focus on competition, organisers believe that the slam nature of Science Slams is of central importance. They contend that audiences at slams enjoy judging the slammers, and that people are more likely to attend academic talks when they are allowed to judge performances. Although the audience may feel empowered by this practice, many slammers, who enjoy the 'community building' and 'networking' aspect of the event, dislike being judged by the audience.

The New Construction of the Public

If we apply Berger and Luckmann's 'bigger stick' theory to knowledge circulation in the public, it would most likely be acknowledged that the people who hold the biggest stick nowadays are those who dominate the internet.[7] When considering the production of the public sphere and knowledge circulation today, the digitalisation of society is of importance. The social dissemination of the Science Slam in Germany was partially based on the organisers' practice of uploading videos of slams to YouTube, which started in 2009. With the rise of digital platforms, we can also observe a strong shift in our capitalistic structures. Platforms are 'a discrete and dynamic arrangement defined by particular combination of socio-technical and capitalist business practices' (Langley and Leyshon 2016, 13). According to Langley and Leyshon, platforms are based on user participation and are oriented towards venture capital funds. The digitalisation of the circular economy occurred when new business models were developed through the digitalisation of information and communication technologies, which broke down traditional market structures. Multifaceted platforms enable new exchange-based relationships, for example, online exchange markets, social media and user-generated content, sharing economy, crowdsourcing and crowdfunding. Take the example of YouTube, which was founded in the United States, in 2005.[8] YouTube enables users to upload videos and makes most of its money through advertisements (it made $15 billion in 2019).[9] Other platforms make money through predicting user behaviour, which they then sell to their customers (Zuboff 2019).[10] As these new online companies give rise to precarious employment and unsustainable business practices (Irani 2015), many scholars ask us to reflect critically on them. Of most concern is how platforms polarise public discourse by producing 'traffic' based on unfounded knowledge, conspiracy theories, and propaganda.

In the past few decades, before internet platforms were invented, the public's political opinions were hugely influenced by forms of mass media because they offered a critical reflection on political activities. With the invent of social media, classical mass media has a competitor. Internet platforms and social media companies do not think that investigating politics is their job and they do not wish to be 'gatekeepers' of information, having to fact-check and release or share factual, researched knowledge only. Their business models are built on generating as much traffic through their websites as possible and since 'fake news' and conspiracy theories are useful in generating traffic and allow internet companies to sell more user data, they do not have a motive for stopping it. On the plus side, social media allows users to experience 'individualised possibility of constant and disruptive intervention in political events' (Hofman 2018, 17), which creates many opportunities for a revival of politics (Arab Spring, Fridays for Future, Black Lives Matter) and vibrant democracy.

Scholars contemplate on the relationship between democracy and digitisation. Within this contemplation, the internet is regularly portrayed as both a driving and a hindering force in society and democracy. Some scholars disagree, and argue that democracy is not static and that the internet is not a purely self-legislated and linearly evolving thing, but is characterised by developmental openness (c.f. Hofman 2018, 14). I agree with this and contend that democracy and technical environments are socially produced. Even so, it is also clear that we can lose control of both. When democratic parties can no longer interest an increasingly apathetic mass, the mass can be drawn in to populist movements. Within these populist movements, where totalitarian worldviews are accepted, a false ideology is spread through fear, and nonfactual, unresearched information from the internet is trusted completely. The feeling of needlessness, which we explored earlier in Arendt's work, has most likely increased in societies since Covid-19 hit the world in 2020. With Arendt's analysis of historical totalitarianism in mind, this may lead to a situation where democracy has to prevent a combination of leader principle, a modern bureaucratic mass society, and terror.[11] Our dystopia would be a totalitarian regime in the control of the internet and the public sphere.[12] It is our responsibility to improve our democracies and the digital machine. Democracy and technical environments are socially produced, so we should reconstruct them in order to retain freedom within our society.

In digital societies, social action is distributed to a greater extent between human behaviour and technical processes. Technical agencies are given a varying degree of importance and attention within sociological debates on technology (Rammert 1993) and debates within science and technology studies, for example, the 'Actor Network Theory' (Latour 2008). When dealing with information systems, problems surrounding morality are exacerbated, because in discussions about codified ethics questions about responsibility arise. Broadly speaking, technology encompasses devices and generally formulates to embody purposeful means.[13] What internet users view as personalised information streams are filtered and hierarchised by algorithms. Algorithms are not neutral instruments but have values inscribed into their functional logic, which has an impact on public discourse. They have a conceptual sequence of steps and they solve a problem based on a set of rules. We live in a time where the ideal, in science at least, is mechanical objectivity (Daston and Galison 2007), which aims to eliminate all human interference, either through machines or through the mechanisation of scientific procedures. At first it seems as though algorithms are the perfect realisation of mechanical objectivity. Algorithms are designed by humans, but the algorithm acts autonomously (based on machine learning). It recognises patterns, reads traces, and develops strategies on its own (Cardon 2017). However, when algorithms are used for internet platforms they are often used for economic purposes only, so they reflect the economic interests of their constructors. Therefore, algorithms are not the perfect realisation of mechanical objectivity because normativity and values like racism are written into algorithms (Noble 2018).[14] Additionally, algorithms can develop independency, and the loss of control over these systems causes fake news to spread extremely fast in our society. Consequently, Zittrain labels algorithms 'the new gatekeepers of the digital public sphere' (Zittrain quoted in Cardon 2017, 1).

In recent years, public spheres have changed. Researchers today talk of algorithmically generated personalised publics (Hofman 2018, 17), where it is difficult to talk about shared reality and shared knowledge, because we do not know what others see. The 'alternative facts' which many people believe often become independent ideas once they are shown on social media channels or on private TV channels, which results in them becoming an alternative reality for some. This alternative realities for some have the result that we meet as strangers. In his writings, Schütz (1972) uses the example of the stranger to make clear what it means when one arrives in a new environment without trustworthy recipes. If one comes into a society as a stranger, one does not have these recipe systems as interpretation and instruction schemes, and is therefore fundamentally insecure. The stranger must translate expressions into those of the civilisation patterns of his home group, provided that an interpretative equivalent exists at all within the latter (cf. ibid., 63). A foreigner cannot take for granted what is self-evident, and cannot accept unquestioned what is unquestioned. He must laboriously learn the typification and relevancies of the new society if he is to adapt socially (cf. ibid., 70). People might feel like strangers in new digitalised worlds. In my research, I noticed Haraway's (1997) concept of a cyborg. The interweaving of people and technology was evident on stage when slammers interacted with their slides, it was evident in the audience when waiting guests constantly looked at their smartphones, and it was evident in the development of the genre as such when new performances were inspired by old successful performances from the YouTube archive. I contend that current public spheres cannot be compared to the ancient Greek idea of the polis, where people would meet in the city to debate. Nowadays, people are deeply intertwined with their technological devices (Kaerlein 2018; Hepp 2020) and the digital world, which they rely on to organise any meetings in a public space. We are witness to a subjectivation and emerging publics in a back-and-forth movement between human and machine (Asenbaum 2018).[15] In this context many are immature citizens. Close devices seem fairly strange. They do not understand the technical environment that they have merged with and they do not understand other mysteriously emerged counterpublics. Self-inflicted immaturity in the sense of the Enlightenment today means not knowing the backstage of the digital world and blindly trusting information on the net. In a world of cyborgs in cyborg public spheres, we may have to explore other novel forms of public participation and communication (Hepp 2016). People seem to be lost in translation. This is of course no excuse for counter-publics based on conspiracy theories.

Throughout this book I have analysed new modes of production in public discourse, specifically related to science and the public, within the context of a digital society. Societies today are partly defined by internet platforms, and partly by particular combinations of socio-technical and capitalist business practices. In the case of the Science Slam, new reflexive modes of genre production have been established between YouTube and public performances on stage. An algorithmically generated and personalised archive has influenced how the public are performantly produced in co-presence. This digital sphere has influenced ongoing public formation. This 'reflexive production' mode would have an impact if YouTube's algorithms suggested other Science Slam videos for users to watch. Algorithmically generated, personalised information has had an impact on society and on the further development of genres. I predict that in the next few years we will see how important the 'truth' is to network operators and owners of internet platforms.

Notes

1 It could be suggested that contemporary scientists are not that privileged due to short contracts, underpayment, and their uncertain futures. Yet, when we consider their relation to society as a whole, I believe they should be considered so.

2 For example, the call for scientists to provoke astonishment is an example of the genre's novelty. However, this same feature was present in the classic 17th century's skilful demonstrations of scientific principles by scientists for visitors and witnesses.

3 The new scientific persona may now include a young woman (Greta Thunberg).

4 This part of the text has already been published (Wilke and Hill 2019).

5 Organisers say that the need for scientists to communicate with the public, or non-specialists, has grown. This belief in disseminating scientific knowledge is rooted in a commitment to the scientist as a single author (as we saw when Science Slam organisers discussed the importance of having self-produced knowledge in slams). The authors of knowledge have become significantly more important in public science communication.

6 This idea of a community was highlighted by several slammers I interviewed.

7 'Surveillance capitalism operates through unprecedented asymmetries in knowledge and the power that accrues to knowledge. Surveillance capitalists know everything about us, whereas their operations are designed to be unknowable to us. They accumulate vast domains of new knowledge from us, but not for us. They predict our futures for the sake of other's gain, not ours' (Zuboff 2019, 11).

8 Its slogan is 'Broadcast yourself'.

9 Accessed February 2, 2021, www.faz.net/aktuell/wirtschaft/digitec/google-zahlen-fuer-alphabet-viertes-quartal-2019-16616065.html.

10 'Surveillance claims human experience as free raw material for translation into behavioural data. Although some of these data are applied to service improvement, the rest are declared as a proprietary behavioural surplus, fed into advanced manufacturing processes known as "machine intelligence", and fabricated into prediction products that anticipate what you will do now, soon, and later. Finally, these prediction products are traded in a new kind of marketplace that I call behavioural futures markets. Surveillance capitalists have grown immensely wealthy from these trading operations, for many companies are willing to lay bets on our future behaviour' (Zuboff 2019, 18).

11 In agreement with Arendt, I believe no one should be allowed to argue that they are just 'a cog in the machine'. When Mark Zuckerberg talked to the US congress in a hearing about Facebook's influence on the 2016 US Election, it was a strong reminder of the organised irresponsibility that Arendt warned about. In general, designers of internet platforms should not control a technical and bureaucratic apparatus without being held responsible for the actions that result from or through these.

12 Consequently, it is not only Trump who should be held responsible for the actions taken during his administration. It is not acceptable that Republicans, Fox News, and

other social media platforms bear no responsibility for their part in aiding and allowing the administration to carry out some of its more problematic actions. It is not acceptable that violence against black people is allowed to continue and is not stopped by the government and it is not acceptable that social media platforms spread false information without facing any consequences. Legal order and state regulations must catch up with new developments. Rules that apply to the economic sphere should be applied to internet platforms, while rules that apply to traditional media forms should be applied to digital media forms.

13 'Technology is usually defined as tools made by man, as efficient means to an end, or as an ensemble of material artifacts. But technology also encompasses instrumental practices, like the creation, fabrication and the use of means and machines; it includes the whole ensemble of material and non-material techno-facts; it is closely connected with institutionalized needs and ends-in-view that technologies serve to'. Werner Rammert, accessed February 13, 2021, www.ts.tu-berlin.de/fileadmin/fg226/Rammert/articles/Relation.html.

14 This demonstrates how social inequalities are produced through the internet. Noble argues, therefore, that mechanical, computer-based calculations must be accompanied by science and social science which can critique them and explain what consequences this information may result in.

15 Like Haraway's cyborg metaphor, 'cyborg publics' acknowledges the blurred boundary between bodies (organism) and tools (machines). This is particularly interesting because people seem to perceive smartphones as part of their body (Kaerlein 2018).

References

Abbott, Andrew. 2001. *Chaos of Disciplines*. Chicago: University of Chicago Press.

Abu-Lughod, Lila. 1990. "Can There Be a Feminist Ethnography?" *Woman and Performance: A Journal of Feminist Theory* 5(1): 7–27.

Adorno, Theodor W., and Max Horkheimer. 2002. *Dialectic of Enlightenment*. Stanford: Stanford University Press.

Alac, Morana. 2008. "Working with Brain Scans: Digital Images and Gestural Interaction in fMRI Laboratory." *Social Studies of Science* 38(4): 483–508.

Amann, Klaus, and Karin Knorr-Cetina. 1990a. "The Fixation of (Visual) Evidence." In *Representations in Scientific Practice*, edited by Michael Lynch and Steve Woolgar, 86–121. Cambridge: MIT Press.

———. 1990b. "Image Dissection in Natural Scientific Inquiry." *Science, Technology & Human Values* 15(3): 259–83.

Anders, Petra. 2008. *Texte und Materialien für den Unterricht: Slam Poetry*. Stuttgart: Reclam.

Arendt, Hannah. 1960. *Vita Activa oder vom tätigen Leben*. Stuttgart: Kohlkammer.

———. 1965. *Über die Revolution*. München: Piper.

———. 1970a. *Macht und Gewalt*. München: Piper.

———. 1970b. *On Violence*. New York: Houghton Mifflin Harcourt.

———. 2006. *Über das Böse: Eine Vorlesung zu Fragen der Ethik*. Text of the Lecture. München: Piper.

———. 2017. *Elemente und Ursprünge totalitärer Herrschaft: Antisemitismus, Imperialismus, totale Herrschaft*. Frankfurt a.M: Europäische Verlagsanstalt.

Asenbaum, H. 2018. "Cyborg Activism: Exploring the Reconfigurations of Democratic Subjectivity in Anonymous." *New Media & Society* 20(4): 1543–63.

Aspers, Patrik. 2014. "Performing Ontology." *Social Studies of Science* 45(3): 449–53.

Austin, John L. 1975. *How to Do Things with Words: The William James Lectures Delivered at Harvard University in 1955*. Cambridge: Harvard University Press.

Barad, Karen. 2003. "Posthumanist Performativity: Toward an Understanding of How Matter Comes to Matter." *Signs: Journal of Women in Culture and Society* 28(3), Spring.

Barnes, Barry. 1974. *Scientific Knowledge and Sociological Theory*. London: Routledge.

Bauernschmidt, Stefan. 2018. "Öffentliche Wissenschaft, Wissenschaftskommunikation & Co.: Zur Kartierung zentraler Begriffe in der Wissenschaftskommunikationswissenschaft." In *Öffentliche Gesellschaftswissenschaften zwischen Kommunikation und Dialog*, hrsg by Stefan Selke and Annette Treibel. Wiesbaden: Springer.

Beaufays, Sandra, and Beate Krais. 2005. "Doing Science—Doing Gender: Die Produktion von WissenschaftlerInnen und die Reproduktion von Machtverhältnissen." *Feministische Studien* 23(1): 82–99.

Beaulieu, Anne. 2002. "Images Are Not the (Only) Truth: Brain Mapping, Visual Knowledge, and Iconoclasm." *Science, Technology & Human Values* 27(1): 53–83.

Beck, Gerald. 2013. *Sichtbare Soziologie: Visualisierung und Wissenschaftskommunikation in der Zweiten Moderne*. Bielefeld, Germany: Transcript.

Beck, Ulrich. 1986. *Risikogesellschaft: Auf dem Weg in eine andere Moderne*. Frankfurt a.M.: Suhrkamp.

Beer, David. 2016. *Metric Power*. London: Palgrave Macmillan.

Berger, Peter L., and Thomas Luckmann. 1967. *The Social Construction of Reality*. New York: Penguin Books.

Bloor, David. 1976. *Knowledge and Social Imagery*. Chicago: University of Chicago Press.

Böhme, Gernot, and Nico Stehr, eds. 1986. *The Knowledge Society: The Growing Impact of Scientific Knowledge on Social Relations*. Dodrecht: Reidel.

Born, Dorothea. 2015. "Communicating Science, Transforming Knowledge: Insights into the Production Processes of the Popular Science Magazine GEO." In *Studying Science Communication: Postgraduate Papers*, edited by Erik Stengler. Bristol: University of West England.

Bourdieu, Pierre. 1988. *Homo academicus*. Frankfurt a.M.: Suhrkamp.

Braun-Thürmann, Holger. 2005. *Innovation*. Bielefeld: Transcript-Verlag.

Bröckling, Ulrich. 2007. *Das unternehmerische Selbst: Soziologie einer Subjektivierungsform*. Frankfurt a.M.: Suhrkamp.

Bucchi, Massimiano. 2008. "Of Deficits, Deviations and Dialogues: Theories of Public Communication of Science." In *Handbook of Public Communication of Science and Technology*, edited by Massimiano Bucchi and Brian Trench, 57–76. London: Routledge.

Bucchi, Massimiano, and Brian Trench, eds. 2008. *Handbook of Public Communication of Science and Technology*. London: Routledge.

Bucher, Tania. 2012. "Want to Be on Top? Algorithmic Power and the Threat of Visibility on Facebook." *New Media& Society* 14(7): 1164–80.

Bultude, Karen, Dominic McDonald, and Savita Custead. 2011. "The Rise and Rise of Science Festivals: An International Review of Organised Events to Celebrate Science." *International Journal of Science Education, Part B: Communication and Public Engagement* 1(2): 165–88.

Büscher, M. 2005. "Social Life Under the Microscope?" *Sociological Research Online* 10(1): 100–23. doi:10.5153/sro.966

Butler, Judith. 1998. *Haß spricht: Zur Politik des Performativen*. Berlin: Berlin Verlag.

———. 2002. "Performative Akte und Geschlechterkonstitution: Phänomenologie und feministische Theorie." In *Performanz: Zwischen Sprachphilosophie und Kulturwissenschaften*, edited by Uwe Wirth, 301–20. Frankfurt a.M.: Suhrkamp.

Cardon, Dominique. 2017. "Den Algorithmus dekonstruieren: Vier Typen digitaler Informationsberechnung." In *Algorithmuskulturen: Über die rechnerische Konstruktion der Wirklichkeit*, edited by Robert Seyfert and Jonathan Roberge, 131–50. Bielefeld: Transcript.

Chomsky, Noam. 1981. *Regeln und Repräsentationen*. Frankfurt a.M.: Suhrkamp.

Collins, Harry. 1983. "An Empirical Relativist Programme in the Sociology of Scientific Knowledge." In *Science Observed: Contemporary Analytical Perspectives*, edited by Karin Knorr-Cetina and Joseph Mulkay, 85–114. London: Sage.

———. 1987. "Certainty and the Public Understanding of Science: Science on Television." *Social Studies of Science* 17(4): 684–713.

Collins, Harry, and Robert Evans. 2002. "The Third Wave of Science Studies: Studies of Expertise and Experience." *Social Studies of Science* 32(2): 235–96.

Collins, Harry, and Trevor Pinch. 1993. *The Golem: What Everyone Should Know About Science*. Cambridge: Cambridge University Press.

Crouch, Colin. 2004. *Post-Democracy*. Oxford: Oxford University Press.

Daston, Lorraine. 2001. "Die Kultur der wissenschaftlichen Objektivität." In *Ansichten der Wissenschaftsgeschichte*, hrsg by Michael Hagner, 137–60. Frankfurt a.M: De Gruyter.

———. 2003. "Die wissenschaftliche Persona: Arbeit und Berufung." In *Zwischen Vorderbühne und Hinterbühne: Beiträge zum Wandel der Geschlechterbeziehungen in der Wissenschaft vom 17. Jahrhundert bis zur Gegenwart*, edited by Theresa Wobbe Wobbe, 109–36. Bielefeld: Transcript.

Daston, Lorraine, and Peter Galison. 2007. *Objectivity*. New York: Zone Books.

Daston, Lorraine, and Katharine Park, eds. 2006. *The Cambridge History of Science Vol. III: Early Modern Science*. Cambridge: Cambridge University Press.

Daum, Andreas. 1998. *Wissenschaftspopularisierung im 19. Jahrhundert. Bürgerliche Kultur, naturwissenschaftliche Bildung und die deutsche Öffentlichkeit 1848–1914*. München: Oldenbourg.

Davies, Sarah R. 2009. "Doing Dialogue: Genre and Flexibility in Public Engagement with Science." *Science as Culture* 18(4): 397–416.

———. 2013. "Constituting Public Engagement Meanings and Genealogies of PEST in Two U.K. Studies." *Science Communication* 35(6): 687–707.

———. 2014. "Knowing and Loving: Public Engagement Beyond Discourse." *Science & Technology Studies* 27(3): 90–110.

Davies, Sarah R., Elllen Mccallie, Elin Simonsson, Jane Lehr, and Sally Duensing. 2009. "Discussing Dialogue: Perspectives on the Value of Science Dialogue Events That Do Not Inform Policy." *Public Understanding of Science* 18(3): 338–53.

De Beauvoir, Simone. 1949 [1974]. *The Second Sex*. Edited by H. M. Parshley. New York: Knopf.

Dijkstra, Anne M., and Christine R. Critchley. 2014. "Nanotechnology in Dutch Science Cafés: Public Risk Perceptions Contextualised." *Public Understanding of Science* 71–87.

Durkheim, Emile. 1981. *Die elementaren Formen des religiösen Lebens*. Frankfurt a.M.: Suhrkamp.

Ebel, Gerhard, and Otto Lührs. 1988. "Urania: Eine Idee, eine Bewegung, eine Institution wird 100 Jahre alt!" In *100 Jahre Urania Berlin, in: Festschrift Wissenschaft heute für Morgen*, 15–74. Berlin: Urania.

Ebner, Julia. 2019. *Radikalisierungsmaschinen*. Frankfurt a.M.: Suhrkamp.

Elias, Norbert. 1969. *Über den Prozeß der Zivilisation: Soziogenetische und psychogenetische Untersuchungen*. Frankfurt a.M.: Suhrkamp.

Etzkowitz, Henry. 1998. "The Norms of Entrepreneurial Science: Cognitive Effects of the New University-Industry Linkages." *Research Policy* 27(8): 823–33.

Etzkowitz, Henry, and Loet Leydesdorff. 2000. "The Dynamics of Innovation: From National Systems and 'Mode 2' to a Triple Helix of University-Industry-Government Relations." *Research Policy* 29(2): 109–23.

Eyal, Gil, and Larissa Buchholz. 2010. "From the Sociology of Intellectuals to the Sociology of Interventions." *Annual Review of Sociology* 36: 117–37.

Findlen, Paula. 1999. "Masculine Prerogatives: Gender, Space, and Knowledge in the Early Modern Museum." In *Architecture of Science*, edited by Peter Galison and Emily Thompson, 29–57. Boston: MIT Press.

Fischer, Georg, and Lorenz Grünewald-Schukalla. 2018. "Originalität und Viralität von (Internet) Memes." *Sonderausgabe kommunikation@gesellschaft* 19.

Fischer-Lichte, Erika, ed. 2001. *Theatralität und die Krisen der Repräsentation: DFG-Symposion 1999*. Stuttgart: JB Metzler.

———. 2011. *Performativität: Eine Einführung*. Bielefeld: Transcript.

Fischer-Lichte, Erika, Christian Horn, Sandra Umathum, and Matthias Warsta, eds. 2004. *Theatralität als Modell in den Kulturwissenschaften*. Tübingen und Basel: A Francke.

Fleck, Ludwik. 1980 [1935]. *Entstehung und Entwicklung einer wissenschaftlichen Tatsache: Einführung in die Lehre vom Denkstil und Denkkollektiv*. Frankfurt a.M.: Suhrkamp.

Flick, Uwe. 2005. "Wissenschaftstheorie und das Verhältnis von qualitativer und quantitativer Forschung." In *Qualitative Medienforschung: Ein Handbuch*, edited by Lothar Mikos and Claudia Wegener, 20–28. Konstanz: UVK.

Foucault, Michel. 1972. *The Archaeology of Knowledge and the Discourse on Language*. New York: Pantheon.

Fraser, Nancy. 1992. "Rethinking the Public Sphere: A Contribution to the Critique of Actually Existing Democracy." In *Habermas and the Public Sphere*, hrg. by Craig Calhoun, 109–42. Cambridge, MA and London: MIT Press.

Freter, Kristin. 2019. "Macht- und Gewaltstrukturen im Sprachduktus der Neuen Rechten." *Macht und Gewalt* 159–81.

Garfinkel, Harold. 1967. *Studies in Ethnomethodology*. Englewood Cliffs, NJ: Prentice Hall.

Gibbons, Michael, Camille Limoges, Helga Nowotny, Simon Schwartzman, Peter Scott, and Martin Trow. 1994. *The New Production of Knowledge: The Dynamics of Science and Research in Contemporary Societies*. London: Sage.

Giddens, Anthony. 1990. *The Consequences of Modernity*. Cambridge: Cambridge Polity Press.

Gieryn, Thomas F. 1983. "Boundary-Work and the Demarcation of Science from Non-Science: Strains and Interests in Professional Ideologies of Scientists." *American Sociological Review* 48(6): 781–95.

Gilfillan, Seabury C. 1935 [1970]. *The Sociology of Invention: An Essay in the Social Causes, Ways and Effects of Technic Invention, Especially, as Demonstrated Historically in the Author's 'Inventing the Ship'*. Cambridge, MA: MIT Press.

Godin, Benoît. 2008. "In the Shadow of Schumpeter: W. Rupert Maclaurin and the Study of Technological Innovation." *Minerva* 46(3): 343–60.

———. 2010. "Innovation Without the Word: William F. Ogburn's Contribution to the Study of Technological Innovation." *Minerva* 48(3): 277–307.

———. 2014a. *Innovation and Science: When Science Had Nothing to Do with Innovation, and Vice-Versa*. Project on the Intellectual History of innovation, Working Paper no. 16. Montreal: INRS.

———. 2014b. *Reimagining Innovation in the Nineteenth Century: A Study in the Rhetoric of Innovation, Project on the Intellectual History of Innovation*. Montreal: INRS.

———. 2015. *Innovation Contested: The Idea of Innovation Over the Centuries*. London: Routledge.

Goffman, Erving. 1953. "Communication Conduct in an Island Community" (Unpublished PhD thesis, Department of Sociology, University of Chicago, Chicago).

————. 1959. *The Presentation of Self in Everyday Life*. New York: Anchor Books.

————. 1974. *Frame Analysis: An Essay on the Organization of Experience*. Boston: Northeastern University Press.

————. 1981. *Forms of Talk*. Philadelphia, PA: University of Pennsylvania Press.

Goodwin, Charles. 1981. *Conversational Organization: Interaction Between Speakers and Hearers*. New York: Academic.

————. 1994. "Professional Vision." *American Anthropologist* 96(3): 606–33.

Greco, Monica, and Paul Stenner. 2008. "Introduction: Emotion and Social Science." In *Emotions: A Social Science Reader*, edited by Monica Greco and Paul Stenner, 1–22. London: Routledge.

Gregory, Jane, and Steve Miller. 2000. *Science in Public: Communication, Culture, and Credibility*. London: Basic Books.

Groys, Boris. 2004. *Über das Neue*. Versuch einer Kulturökonomie 3 Auflage. Frankfurt am Main: Suhrkamp.

Günthner, Susanne, and Hubert Knoblauch. 1994. "Forms Are the Food of Faith: Gattungen als Muster kommunikativen Handelns." *KZfSS* 46(4): 693–723.

————. 2007. "Wissenschaftliche Diskursgattungen—PowerPoint et al." In *Reden und Schreiben in der Wissenschaft*, edited by Peter Auer and Harald Baßler, 53–65. Frankfurt a.M.: Campus.

Habermas, Jürgen. 1990. *Strukturwandel der Öffentlichkeit*. Frankfurt a.M.: Suhrkamp.

Hacking, Ian. 1999. *The Social Construction of What?* Cambridge: Harvard University Press.

Haraway, Donna. 1988. "Situated Knowledges: The Science Question in Feminism and the Privilege of Partial Perspective." *Feminist Studies* 14(3): 575–99.

————. 1989. *Primate Visions: Gender, Race, and Nature in the World of Modern Science*. New York: Routledge.

————. 1991. "A Cyborg Manifesto: Science, Technology and Socialist Feminism in the Late Twentieth Century." In *Simians, Cyborgs, and Women: The Reinvention of Nature*. London: Free Association Books.

————. 1997. *Modest_Witness@Second_Millennium: FemaleMan©_Meets_OncoMouse*[a]: *Feminism and Technoscience*. London: Routledge.

Harding, Sandra. 1986. *The Science Question in Feminism*. Milton Keynes: Open University Press.

Haynes, Roslynn D. 1994. *From Faust to Strangelove: Representations of the Scientist in Western Literature*. Baltimore: Johns Hopkins University Press.

Hegel, Georg Wilhelm Friedrich. 1955. *Grundlinien der Philosophie des Rechts*. Hamburg: Meiner.

Heintz, Bettina, Martina Merz, and Schumacher Christina, eds. 2004. *Wissenschaft, die Grenzen schafft: Geschlechterkonstellationen im disziplinären Vergleich*. Bielefeld: Transcript.

Hepp, Andreas. 2016. "Pioneer Communities." *Media, Culture & Society* 38(6): 918–33.

————. 2020. *Deep Mediatization*. London: Routledge.

Herbrik, Regine. 2011. *Die kommunikative Konstruktion imaginärer Welten*. Wiesbaden: VS.

Hettling, Manfred, and Bernd Ulrich, eds. 2005. *Bürgertum nach 1945*. Hamburg: Hamburger Edition.

Hildenbrand, Bruno. 2009. "Anselm Strauss." In *Qualitative Forschung: Ein Handbuch*, edited by Uwe Flick, Ernst von Kardorff, and Ines Steinke, 32–41. Reinbek: Rowohlt.

Hill, Miira. 2015. "Science Slam und die Darstellung von,Tatsachen'- eine Vergessenheit der Wissensproduktion?." In *Auf der Suche nach den Tatsachen: Proceedings der 1*, hrsg by J. Engelschalt and A. Maibaum, 127–41. Berlin: Tagung des Nachwuchsnetzwerks INSIST.

———. 2017a. "Die Versinnbildlichung von Gesellschaftswissenschaft—Herausforderung Science Slam." In *Öffentliche Wissenschaft und gesellschaftlicher Wandel*, edited by Stefan Selke und Annette Treibel. Wiesbaden: Springer VS.

———. 2017b. "Science Slam und die (Re)Präsentation von Wissenschaft: Neue Einsichten des Kommunikativen Konstruktivismus über Wissenschaftskommunikation in der Popkultur." In *Der Kommunikative Konstruktivismus bei der Arbeit*, edited by Jo Reichertz und René Tuma. Weinheim and München: Juventa Verlag.

———. 2020. "Innovative Popular Science Communication? Materiality, Aesthetics and Gender in Science Slams." In *Genealogy of Popular Science: From Ancient Ecphrasis to Virtual Reality*, edited by J. M. Morcillo and C. Y. Robertson-von Trotha. Bielefeld: Transcript Verlag.

Hitzler, Ronald. 1998. "Posttraditionale Vergemeinschaftung: Über neue Formen der Sozialbindung." *Berliner Debatte Initial* 9(1): 81–89.

Hofman, Jeanette. 2018. *Digitalisierung und demokratischer Wandel als Spiegelbilder (2018)*, 14–21. Martinsen: Wissen, Macht, Meinung.

Hutter, Michael. 2011. "Infinite Surprises: On the Stabilization of Value in the Creative Industries." In *The Worth of Goods: Valuation and Pricing in the Economy*, edited by Jens Beckert and Patrik Aspers, 201–22. London: Oxford University Press.

———. 2015a. "Zur Rolle des Neuen in der Erlebnisgesellschaft und ihrer Wirtschaft." In *Innovationsgesellschaft heute*, edited by Werner Rammert, Arnold Windeler, Hubert Knoblauch, and Michael Hutter. Wiesbaden: Springer VS.

———. 2015b. *The Rise of the Joyful Economy*. Milton Park: Routledge.

Hutter, Michael, Hubert Knoblauch, Werner Rammert, and Arnold Windeler. 2011. *Innovationsgesellschaft heute: Die reflexive Herstellung des Neuen*. Technical University Technology Studies Working Papers. Berlin: TU Berlin.

Hymes, Dell. 1971. *On Commincative Competence*. Philadelphia: University of Pennsylvania Press.

———. 1974. *Foundations in Sociolinguistics: An Ethnographic Approach*. Philadelphia: University of Pennsylvania Press.

Irani, Lilly. 2015. "Difference and Dependence Among Digital Workers: The Case of Amazon Mechanical Turk." *South Atlantic Quarterly* 114(1): 225–34.

Irwin, Alan. 2001. "Constructing the Scientific Citizen: Science and Democracy in the Biosciences." *Public Understanding of Science* 10(1): 1–18.

Jarausch, Konrad H. 1990. *The Unfree Professions: German Lawyers, Teachers, and Engineers, 1900–1950*. Oxford: Oxford University Press.

Jasanoff, Sheila. 1997. "Civilization and Madness: The Great BSE Scare of 1996." *Public Understanding of Science* 6(3): 221–32.

———. 2003. "Breaking the Waves in Science Studies: Comment on H.M. Collins and Robert Evans, 'The Third Wave of Science Studies'." *Social Studies of Science* 33(3): 389–400.

Joas, Hans. 1992. *Die Kreativität des Handelns*. Frankfurt a.M.: Suhrkamp.

Johns, Adrian. 2006. "Coffeehouses and Print Shops." In *The Cambridge History of Science III: Early Modern Science*, edited by Katharine Park, 320–40. Cambridge: Cambridge University Press.

Kaerlein, Timo. 2018. *Smartphones als digitale Nahkörpertechnologien—Zur Kyber-netisierung des Alltags*, 35–91. Bielfeld: Transcript.

Kelle, U., and C. Erzberger. 2005. "Qualitative und quantitative Methoden: Kein Gegensatz." In *Qualitative Forschung*, edited by U. Flick, E. von Kardoff, and I. Steinke. Reinbek bei Hamburg: Rowohlt.

Keller, Reiner, Hubert Knoblauch, and Jo Reichertz, eds. 2013. *Kommunikativer Konstruktivismus: Theoretische und empirische Arbeiten zu einem neuen wissenssoziologischen Ansatz*. Wiesbaden: Springer VS.

Kiesow, Christian. 2014. *Exotische Sphären—eine wissenssoziologische Studie zu Kommunikation, Interaktion und Visualität in der mathematischen Forschung*. Berlin: Technische Universität.

Knoblauch, Hubert. 1995. *Kommunikationskultur: Die kommunikative Konstruktion kultureller Kontexte*. Berlin: De Gruyter.

———. 2001. "Fokussierte Ethnographie." *Sozialer Sinn Heft* 2(1): 123–41.

———. 2003. *Qualitative Religionsforschung: Religionsethnographie in der eigenen Gesellschaft*. Paderborn: UTB.

———. 2005a. *Wissenssoziologie*. Konstanz: UVK Verlag.

———. 2005b. "Kulturkörper: Die Bedeutung des Körpers in der sozialkonstruktivistischen Wissenssoziologie." In *Soziologie des Körpers*, edited by Markus Schroer, 92–113. Frankfurt a.M.: Suhrkamp.

———. 2007. "Die Performanz des Wissens: Zeigen und Wissen in der Powerpoint-Präsentation." In *Powerpoint-Präsentationen: Neue Formen der gesellschaftlichen Kommunikation von Wissen*, edited by Bernt Schnettler and Hubert Knoblauch, 117–38. Konstanz: UVK.

———. 2008a. "Kommunikationsgemeinschaften: Überlegungen zur kommunikativen Konstruktion einer Sozialform." In *Posttraditionale Gemeinschaften: Theoretische und ethnografische Erkundungen*, edited by Ronald Hitzler, Anne Honer, and Michaela Pfadenhauer, 73–88. Wiesbaden: Springer VS.

———. 2008b. "The Performance of Knowledge: Pointing and Knowledge in Powerpoint Presentations." *Cultural Sociology* 2(1): 75–97.

———. 2011. "Alfred Schütz, die Phantasie und das Neue: Überlegungen zu einer Theorie kreativen Handelns." In *Die Entdeckung des Neuen: Qualitative Sozialforschung als Hermeneutische Wissenssoziologie*, edited by Norbert Schröer and Oliver Bidlo, 99–116. Wiesbaden: VS Verlag.

———. 2013. "Grundbegriffe und Aufgaben des Kommunikativen Konstruktivismus." In *Kommunikativer Konstruktivismus: Theoretische und empirische Arbeiten zu einem neuen wissenssoziologischen Ansatz*, edited by Reiner Keller, Hubert Knoblauch, and Jo Reichertz, 25–47. Wiesbaden: Springer VS.

———. 2014. "Communication, Culture and Powerpoint." In *Culture, Communication, and Creativity: Reframing the Relations of Media, Knowledge, and Innovation in Society*, edited by Hubert Knoblauch, René Tuma, and Marc Jacobs, 155–76. Frankfurt a.M.: Peter Lang.

———. 2017. "Gesellschaftstheorie." In *Die kommunikative Konstruktion der Wirklichkeit*. Neue Bibliothek der Sozialwissenschaften. Wiesbaden: Springer VS. https://doi.org/10.1007/978-3-658-15218-5_4

Knoblauch, Hubert, and Jürgen Raab. 2001. "Genres and the Aesthetics of Advertisement Spots." In *Verbal Art Across Cultures*, edited by Hubert Knoblauch and Helga Kotthoff, 195–219. Tübingen: Narr.

Knorr-Cetina, Karin. 1984. *Fabrikation von Erkenntnis: Zur Anthropologie der Naturwissenschaft*. Frankfurt a.M.: Suhrkamp.

———. 1988. "Das naturwissenschaftliche Labor als Ort der Verdichtung von Gesellschaft." *Zeitschrift für Soziologie* 17(2): 85–101.

———. 1989. "Spielarten des Konstruktivismus: Einige Notizen und Anmerkungen." *Soziale Welt* 40(1–2): 86–96.

———. 1999. *Epistemic Cultures: How the Science Make Knowledge*. Cambridge: Harvard University Press.

———. 2002. *Wissenskulturen: Ein Vergleich Naturwissenschaftlicher Wissensformen*. Frankfurt a.M.: Suhrkamp.

Krifka, Sabine. 2000a. "Das Bild des Gelehrten." In *Erkenntnis, Erfindung, Konstruktion: Studien zur Bildgeschichte von Naturwissenschaften und Technik vom 16. bis zum 19. Jahrhundert*, edited by Hans Holländer. Berlin: Gebr. Mann.

———. 2000b. "Schauexperiment—Wissenschaft als belehrendes Spektakel." In *Erkenntnis, Erfindung, Konstruktion: Studien zur Bildgeschichte von Naturwissenschaften und Technik vom 16. bis zum 19. Jahrhundert*, edited by Hans Holländer. Berlin: Gebr. Mann.

Kuhn, Thomas S. 1970. *The Structure of Scientific Revolutions*. 2. Auflage. Chicago: University of Chicago Press.

Langley, Paul, and Andrew Leyshon. 2016. "Plattform Capitalism: The Intermediation and Capitalization of Digital Economic Circulation." *Finance and Society* 3(1): 11–31.

Latour, Bruno. 1990. "Drawing Things Together." In *Representaions in Scientific Practice*, edited by Michael Lynch and Steve Woolgar, 19–68. Cambridge: MIT Press.

———. 1993. *The Pasteurization of France*. Cambridge: Harvard University Press.

———. 1999. *Pandora's Hope: Essays on the Reality of Science Studies*. Cambridge: Harvard University Press.

———. 2003. "Why Has Critique Run Out of Steam? From Matters of Fact to Matters of Concern." *Critical Inquiry* 30(2): 225–48.

———. 2008. *Wir sind nie modern gewesen: Versuch einer Symmetrischen Anthropologie*. 1. Aufl. Frankfurt a.M.: Suhrkamp Verlag.

Latour, Bruno, and Steve Woolgar. 1979 [1986]. *Laboratory Life: The Construction of Scientific Facts*. Princeton: Princeton University Press.

Lenoir, Timothy. 1997. *Instituting Science: The Cultural Production of Scientific Disciplines*. Stanford: Stanford University Press.

Loenhoff, Jens. 2011. "Die Objektivität des Sozialen." In *Schlüsselwerke Des Konstruktivismus*, edited by Bernhard Pörksen, 143–59. Wiesbaden: VS Verlag.

Luckmann, Thomas. 1986. "Grundformen der gesellschaftlichen Vermittlung des Wissens: Kommunikative Gattungen." In *Kultur und Gesellschaft. Sonderheft 27 der Kölner Zeitschrift für Soziologie und Sozialpsychologie*, edited by Friedhelm Neidhardt, Rainer M. Lepsius, and Johannes Weiss. Opladen: Westdeutscher Verlag.

———. 1999. "Wirklichkeiten. Individuelle Konstitution Und Gesellschaftliche Konstruktion." In *Hermeneutische Wissenssoziologie: Standpunkte zur Theorie der Interpretation*, edited by Ronald Hitzler, Jo Reichertz, and Norbert Schröer, 17–28. Konstanz: UVK.

Luckmann, Thomas, and Hubert Knoblauch. 2000. "Gattungsanalyse." In *Qualitative Forschung*, hrsg by Uwe Flick, Ernst von Kardorff, and Ines Steinke. Ein Handbuch, 538–46. Reinbek bei Hamburg: Rowohlt.

Luhmann, Niklas. 1989. *Vertrauen: Ein Mechanismus der Reduktion sozialer Komplexität*. Stuttgart: UTB.

———. 1996. *Die Realität der der Massenmedien*. 2. Auflage. Opladen: Westdeutscher Verlag.

Lynch, Michael. 1988. "The Externalized Retina: Selection and Mathematization in the Visual Documentation of Objects in the Life Sciences." *Human Studies* 11(2–3): 201–34.

———. 1993. *Scientific Practice and Ordinary Action*. Princeton: Princeton University Press.

———. 2013. "Ontography: Investigating the Production of Things, Deflating Ontology." *Social Studies of Science* 43(3): 444–62.

Marx, Karl. 1990 [1867]. *Capital, Volume 1: A Critique of Political Economy*. London: Penguin.

Marx, Karl, and Friedrich Engels. 1968. *Werke Band 40*. Berlin: Dietz Verlag.

McLain, Raymond. 1981. "The Postulate of Adequacy: Phenomenological Sociology and the Paradox of Science and Sociality." *Human Studies* 4(2): 105–30. JSTOR.

Mead, George H. 1982 [1934]. *Mind, Self, and Society*. Edited by Charles W. Morris. Chicago: University of Chicago Press.

Mead, Margaret, and Byers Paul. 1968. *The Small Conference: An Innovation in Communication*. Front Cover. Paris: Mouton.

Mergel, Thomas. 2001. "Die Bürgertumsforschung nach fünfzehn Jahren." *Archiv für Sozialgeschichte* 41(1): 515–38.

Merkens, Hans. 2009. "Auswahlverfahren, Sampling, Fallkonstruktion." In *Qualitative Forschung: Ein Handbuch*, edited by Uwe Flick, Ernst von Kardorff, and Ines Steinke, 286–99. Reinbek: Rowohlt.

Merton, Robert K. 1973. *The Sociology of Science: Theoretical and Empirical Investigations*. Chicago: University of Chicago Press.

Miège, Bernard. 1989. *The Capitalization of Cultural Production*. New York: International General.

Mikl-Horke, Gertraude. 2001. *Soziologie. Historischer Kontext und soziologische Theorie-Entwürfe*. 5. Auflage. München: Oldenbourg.

Münch, Richard. 2013. *Academic Capitalism: Universities in the Global Struggle for Excellence*. London: Routledge.

Myers, Natasha. 2008. "Molecular Embodiments and the Body-Work of Modeling in Protein Crystallography." *Social Studies of Science* 38(2).

———. 2012. "Dance Your PhD: Embodied Animations, Body Experiments and the Affective Entanglements of Life Science Research." *Body & Society* 18(1): 151–89.

Neumann, Michael, and John Andreas Fuchs, eds. 2009. *Mythen Europas: Schlüsselfiguren der Imagination*. Regensburg: Pustet.

Noble, Safiya Umoja. 2018. *Algorithms of Oppression: How Search Engines Reinforce Racism*. New York: New York University Press.

Nowotny, Helga, Peter Scott, and Michael Gibbons. 2001. *Re-Thinking Science: Knowledge and the Public in an Age of Uncertainty*. Oxford: Polity Press.

Ogburn, William F. 1922. *Social Change with Respect to Culture and Original Nature*. New York: B.W. Huebsch, Inc

Owen, Richard, Phil Macnaghten, and Jack Stilgoe. 2012. "Responsible Research and Innovation: From Science in Society to Science for Society, with Society." *Science and Public Policy* 39(6): 751–60.

Park, Robert E. 1928. "Human Migration and the Marginal Man." *American Journal of Sociology* 33(6): 881–93.

Peters, Sibylle. 2011. *Der Vortrag als Performance*. Bielefeld: Transcript.

Pfadenauer, Michaela. 2010. "Experten." In *Diven, Hacker, Spekulanten: Sozialfiguren der Gegenwart*, hrsg by Stephan Moebius and Markus Schroer, 98–108. Frankfurt a. M.: Suhrkamp.

Rammert, Werner. 1993. *Technik aus soziologischer Perspektive: Forschungs-stand—Theorieansätze—Fallbeispiele*. Opladen: Westdeutscher Verlag.

Reckwitz, Andreas. 2012. *Die Erfindung der Kreativität: Zum Prozess gesellschaftlicher Ästhetisierung*. Berlin: Suhrkamp.

Reichertz, Jo. 2013. "Grundzüge des Kommunikativen Konstruktivismus." In *Kommunikativer Konstruktivismus: Theoretische und empirische Arbeiten zu einem neuen wissenssoziologischen Ansatz*, edited by Reiner Keller, Hubert Knoblauch, and Jo Reichertz, 49–68. Wiesbaden: Springer VS.

Robertson-von Trotha, Caroline, and Morcillo, Jesús Munioz. 2018. "Öffentliche Wissenschaft: Von ›Scientific Literacy‹ zu ›Participatory Culture‹." In *Öffentliche Gesellschaftswissenschaften zwischen Kommunikation und Dialog*, hrsg by Stefan Selke and Annette Treibel. Wiesbaden: Springer.

Rössner, Lutz. 1992. *Elfenbeinturm und Wissenschaft: Eine kulturphilosophisch-wissenschaftspolitische Studie*. Braunschweig: Technische Universität Braunschweig.

Sacks, Harvey, Emanuel Schegloff, and Gail Jefferson. 1974. "A Simplest Systematics for the Organisation of Turn-Taking in Conversation." *Language* 50(4): 696–735.

Salmela, Mikko, and Christian von Scheve. 2018. "Emotional Dynamics of Backlash Politics." *Humanity & Society* 42(4). *Emotional Dynamics of Right-and Left-Wing Political Populism*, edited by Joel Busher, Philip Giurlando, and Gavin B. Sullivan.

Sautet, Marc. 1995. *Un café pour Socrate: Comment la philosophie peut nous aider à comprendre le monde d'aujourd'hui*. Paris: R. Laffont.

Schnettler, Bernt, and Hubert Knoblauch, eds. 2007. *Powerpoint-Präsentationen: Neue Formen der gesellschaftlichen Kommunikation von Wissen*. Konstanz: UVK Verlag.

Schulze, Gerhard. 1992. *Die Erlebnisgesellschaft: Kultursoziologie der Gegenwart*. Frankfurt a.M.: Campus.

Schumpeter, Joseph. 1934 [2000]. "Entrepreneurship as Innovation." In *Entrepreneurship: The Social Science View*, edited by Richard Swedberg, 51–75. Oxford: Oxford University Press.

———. 2011. *The Theory of Economic Development: An Inquiry into Profits, Capital, Credit, Interest, and the Business Cycle*. New Brunswick: Transaction.

Schütz, Alfred. 1972. "Der Fremde." In *Gesammelte Aufsätze*, edited by Alfred Schütz. Dordrecht: Springer.

———. 2003. "Über die mannigfachen Wirklichkeiten." In *Theorie der Lebenswelt 1: Die pragmatische Schichtung der Lebenswelt*, edited by Martin Endreß and Ilja Srubar, 181–240. Konstanz: UVK.

Schützeichel, Rainer. 2012. "Wissenssoziologie." In *Handbuch Wissenschaftssoziologie*, edited by Sabine Maasen, Mario Kaiser, Martin Reinhart, and Barbara Sutter Sutter, 17–26. Wiesbaden: Springer VS.

Shapin, Steven. 1989. "The Invisible Technician." *American Scientist* 77(6): 554–63.

———. 1990. "Science and the Public." In *Companion to the History of Modern Science*, edited by Robert C. Olby, Geoffrey Cantor, John Christie, and Frenchy J. Hodges, 990–1007. London: Routledge.

———. 1994. *A Social History of Truth; Civility and Science in Seventeenth-Century England*. Chicago: University of Chicago Press.

———. 1995. "Here and Everywhere: Sociology of Scientific Knowledge." *Annual Review of Sociology* 21(1): 289–321.

———. 2006. "The Man of Science." In *The Cambridge History of Science Vol. III: Early Modern Science*, edited by Lorraine Daston and Katharine Park, 179–91. Cambridge: Cambridge University Press.

Shapin, Steven, and Simon Schaffer. 1985. *Leviathan and the Air-Pump: Hobbes, Boyle, and the Experimental Life*. Princeton: Princeton University Press.

Sheffield, Suzanne L. 2004. *Women and Science: Social Impact and Interaction*. Santa Barbara: ABC-CLIO.

Simmel, Georg. 1964. "Faithfulness and Gratitude." In *The Sociology of Georg Simmel*, edited by Kurt H. Wolff, 379–95. New York: Free Press.

Sismondo, Sergio. 2010. *Introduction to Science and Technology Studies*. London: Wiley-Blackwell.

Smith, Wally. 2009. "Theatre of Use: A Frame Analysis of IT Demonstrations." *Social Studies of Science* 39(3): 449–80.

Soeffner, Hans-Georg. 2001. "Authentizitätsfallen und mediale Verspiegelungen. Inszenierungen im 20. Jahrhundert." In *Symposiumsband "Theatralität und die Krisen der Repräsentation"*, edited by Erika Fischer-Lichte, 165–76. Stuttgart: Metzler.

Stimm, Maria. 2011. *Science Slam—ein Brückenschlag zwischen wissenschaftlichem Wissen und Event: Eine ethnographische Erkundung vor Ort*. Berlin: Unveröffentlichte Masterarbeit HU.

Suchman, Lucy A. 2011. "Anthropological Relocations and the Limits of Design." *Annual Review of Anthropology* 40(1): 1–18.

Summer, Joachim. 2008. "Frankenstein und die literarische Figur des verrückten Wissenschaftlers." In *Mythen Europas Schlüsselfiguren der Imagination. Das*, hrsg by B. van Schlun and M. Neumann, 19, 58–79. Jahrhundert, Regensburg: Pustet.

Swales, John. 1990. *Genre Analysis: English in Academic and Research Settings*. Boston: Cambridge University Press.

Traue, Boris. 2013. "Visuelle Diskursanalyse: Ein programmatischer Vorschlag zur Untersuchung von Sicht- und Sagbarkeiten im Medienwandel." *Zeitschrift für Diskursforschung* 1: 117–36.

Tufte, Edward R. 2006. *The Cognitive Style of Power Point: Pitching Out Corrupts Within*. Military Parade. Cheshire: Graphics Press.

Tuma, René. 2012. "The (Re)Construction of Human Conduct: 'Vernacular Video Analysis'." *Qualitative Sociology Review* 8(2): 152–63.

Tuma, René, Bernt Schnettler, and Hubert Knoblauch. 2013. *Videographie: Einführung in die interpretative Videoanalyse sozialer Situationen*. Wiesbaden: Springer VS.

van Dülmen, Richard. 1977. *Die Gesellschaft der Aufklärer: Studien zur bürgerlichen Emanzipation und aufklärerischen Kultur in Deutschland*. Frankfurt a.M.: Fischer.

Warner, Michael. 2002. *Publics and Counterpublics*. Cambridge: Zone Books.

Weber, Max. 1919 [1992]. *Wissenschaft als Beruf*. Tübingen: Mohr Siebeck.

———. 1946. *Essays in Sociology*, 129–56. New York: Oxford University Press.

———. 1976. *Wirtschaft und Gesellschaft: Grundriß der verstehenden Soziologie*. 5. Auflage. Tübingen: Mohr Siebeck.

———. 1977. "Science as a Vocation." In *From Max Weber: Essays in Sociology*, edited by Hedwig Gerth and Freya Mills. New York: Oxford University Press.

Weingart, Peter. 2005. *Die Wissenschaft der Öffentlichkeit: Essays zum Verhältnis von Wissenschaft, Medien und Öffentlichkeit*. Weilerswist: Velbrück Wissenschaft.

Weingart, Peter, and L. Guenther. 2016. "Science Communication and the Issue of Trust." *Journal of Science Communication* 15(5): C01.

Westermayr, Stefanie. 2004. *Poetry Slam in Deutschland: Theorie und Praxis einer multimedialen Kunstform*. Marburg: Tectum.

Wiebicke, Bernhard. 2010. *Melancholiekonzepte der spanischen Renaissance. Die Liebeskrankheit in Fernando de Rojas "Celestina"*. München: GRIN Verlag.

Wilke, R., and Miira Hill. 2019. "On New Forms of Science Communication and Communication in Science: A Videographic Approach to Visuality in Science Slams and Academic Group Talk." *Qualitative Inquiry* 1–16.

Wilke, R., and E. Lettkemann. 2018. "Die Bewältigung interdisziplinärer Wissenskommunikation im Group Talk: Bausteine einer wissenssoziologischen Gattungsanalyse [Coping with Interdisciplinary Knowledge Communication in Group Talk: Elements of a Genre Analysis]." In *Knowledge in Action: Neue Formen der Kommunikation in der Wissensgesellschaft*, edited by E. Lettkemann, R. Wilke, and H. Knoblauch, 73–107. Wiesbaden, Germany: Springer.

Wilke, R., E. Lettkemann, and H. Knoblauch. 2018. "Präsentationales Wissen [Presentational Knowledge]." In *Knowledge in Action: Wissen, Kommunikation und Gesellschaft (Schriften zur Wissenssoziologie)*, edited by E. Lettkemann, R. Wilke, and H. Knoblauch, 239–72. Wiesbaden, Germany: Springer.

Wynne, Brian. 1992. "Misunderstood Misunderstanding: Social Identities and Public Uptake of Science." *Public Understanding of Science* 1(3): 281–304.

Yates, Joanne, and Wanda Orlikowski. 1992. "Genres of Organizational Communication: A Structurational Approach to Studying Communication and Media." *Academy of Management Review* 17(2): 299–326.

———. 1994. "Genre Repertoire: The Structuring of Communicative Practices." *Administrative Science Quarterly* 541–74, is currently published by Johnson.

———. 2007. "Powerpoint-Präsentation und ihre Abkömmlinge: Wie Gattungen das Handeln in Organisationen prägen." In *Powerpoint-Präsentationen: Neue Formen der gesellschaftlichen Kommunikation von Wissen*, edited by Bernt Schnettler and Hubert Knoblauch, 225–48. Konstanz: UVK.

Yearley, Steven. 1994. "Understanding Science from the Perspective of the Sociology of Scientific Knowledge: An Overview." *Public Understanding of Science* 3(3): 245–58.

Ytreberg, Espen. 2009. "The Question of Calculation: Erving Goffman and the Pervasive Planning of Communication." In *The Contemporary Goffman*, edited by Michael Hviid Jacobsen, 293–312. London: Routledge.

Zapf, Wolfgang. 1989. "Über soziale Innovationen." *Soziale Welt* 40(1–2): 170–83.

Zuboff, Shoshana. 2019. *The Age of Surveillance Capitalism: The Fight for a Human Future at the New Frontier of Power*. New York: Public Affairs.

Zuboff, Shoshana, and Karin Schwandt. 2019. *The Age of Surveillance Capitalism: The Fight for a Human Future at the New Frontier of Power*. London: Profile Books.

Index

For Product Safety Concerns and Information please contact our EU
representative GPSR@taylorandfrancis.com
Taylor & Francis Verlag GmbH, Kaufingerstraße 24, 80331 München, Germany